The Only Electroculture Practical Guide for Beginners

Unlock the Secrets to Faster Plant Growth, Bigger Yields, and Superior Crops Using Coil Coppers, Magnetic Antennas, Pyramids, and More

Written by **JOHN FORD**

© Copyright 2023 by John Ford - All rights reserved.

The content contained within this book may not be reproduced, duplicated or transmitted without direct written permission from the author or the publisher.

Under no circumstances will any blame or legal responsibility be held against the publisher, or author, for any damages, reparation, or monetary loss due to the information contained within this book, either directly or indirectly.

Legal notice:

This book is copyright protected. it is only for personal use. you cannot amend, distribute, sell, use, quote or paraphrase any part, or the content within this book, without the consent of the author or publisher.

Disclaimer notice:

Please note the information contained within this document is for educational and entertainment purposes only. all effort has been executed to present accurate, up to date, reliable, complete information. no warranties of any kind are declared or implied. readers acknowledge that the author is not engaged in the rendering of legal, financial, medical or professional advice. the content within this book has been derived from various sources. please consult a licensed professional before attempting any techniques outlined in this book.

By reading this document, the reader agrees that under no circumstances is the author responsible for any losses, direct or indirect, that are incurred as a result of the use of the information contained within this document, including, but not limited to, errors, omissions, or inaccuracies.

TABLE OF CONTENTS

MY JOURNEY TO ELECTROCULTURE 5

INTRODUCTION 7

CHAPTER 1 - A JOURNEY THROUGH HISTORY: ELECTROCULTURE OVER THE YEARS 9
- Definition of Electroculture 9
- Early Concepts and Experiments 11
- Notable Figures and Innovations 13
- Recent Developments 15

CHAPTER 2 - DEMYSTIFYING ELECTROCULTURE: THE SCIENCE BEHIND THE PROCESS 19
- Introduction to Electricity and Its Generation 20
- Role of Electricity In Nature 22
- Explanation of The Earth Battery Experiment 24

CHAPTER 3 - SOIL: THE FOUNDATION OF AGRICULTURE 27
- Characteristics of Fertile Soil 28
- Importance of Soil Health In Electroculture 30

CHAPTER 4: BENEFITS OF ELECTROCULTURE 33
- Faster Plant Growth 34
- Healthier Plants And Increased Flavor 35
- Higher Crop Yields 37
- Other Benefits Of Electroculture 39

CHAPTER 5 - THE HEART OF THE MATTER: UNDERSTANDING ELECTROCULTURE SYSTEMS 41
- The Basic of Electroculture Systems 42
- Types of electroculture systems 43
- Setting Up an Electroculture System 50
- Maintaining an Electroculture System 53

CHAPTER 6 - GETTING HANDS-ON: TOOLS AND MATERIALS FOR ELECTROCULTURE 55

- COMMON TOOLS 56
- MATERIALS AND DIY INSTRUCTIONS FOR CREATING YOUR OWN TOOLS 57
  - SPIRAL 58
  - ENERGY TOWER / IRISH BASALT TOWER 65
  - LAKHOVSKY RINGS 67
  - PYRAMID 72
  - MAGNETIC ANTENNAS AND CYLINDERS 77
  - GENESA CRYSTAL 81
  - COPPER COIL ANTENNAS 83
- CHOOSING THE RIGHT TOOL FOR YOU 86
- SAFETY PRECAUTIONS AND BEST PRACTICES 88
- CLOCKWISE OR ANTI-CLOCKWISE? 89

CHAPTER 7 - HARNESSING THE POWER OF ELECTROCULTURE: PRACTICAL APPLICATIONS 91

- SOIL MANAGEMENT 92
- SEED GERMINATION 93
- IMPROVING CROP GROWTH AND YIELD 95

CHAPTER 8 – FAQS ABOUT ELECTROCULTURE 97

CHAPTER 9 - GAZING INTO THE CRYSTAL BALL: THE FUTURE OF ELECTROCULTURE 105

- EMERGING TRENDS AND INNOVATIONS 105
- POTENTIAL CHALLENGES AND OPPORTUNITIES 107

CONCLUSION 109

My Journey to Electroculture

Let me take you on a journey, a personal voyage of discovery that I embarked upon a decade ago. My name is John Ford, a humble gardener with a passion for sustainable farming practices. Can you recall a moment when a single article completely changed the course of your life? For me, it was an article about electroculture, the intriguing concept of using electricity to enhance plant growth. As a gardener, I was captivated. As an environmentalist, I was filled with hope.

Imagine me, in the quiet solitude of my backyard garden, experimenting with the principles of electroculture. I started with the simplest of tools, the most basic techniques, and a heart full of curiosity. As I watched my plants flourish, growing faster and healthier, yielding more, and resisting pests and diseases, I felt a sense of awe. It was as if I had unlocked a secret language, a silent dialogue between electricity and life.

But was the path always smooth? Far from it. I stumbled, faced obstacles, and made mistakes. But isn't it through mistakes that we truly learn? Each challenge was a lesson, each setback a stepping stone to deeper understanding. I devoured every piece of research I could find, connected with fellow enthusiasts, and attended conferences. I was a man on a mission, a mission to unravel the mysteries of electroculture.

Over the years, I've witnessed the transformative power of electroculture. I've seen it in the vibrant green leaves of my garden, in the bountiful harvests of farms that have adopted this approach. I've seen it in the way it boosts productivity, enhances plant health, and contributes to sustainable farming. And I've come to believe, with every fiber of my being, that electroculture has the potential to revolutionize the way we grow our food.

This belief, this burning conviction, is what led me to write this book. I wanted to share my journey, to demystify electroculture, and to guide those who are curious about this fascinating field. I wanted to create a beacon, a source of light that could guide beginners and serve as a reference for experienced practitioners.

In this book, I've poured all the knowledge and experience I've gathered over the past decade. I've drawn from my own journey, from the triumphs and the trials, and from the latest research in the field. I've strived to present the information in a way that is clear, concise, and above all, practical. And I've endeavored to convey not just the 'how', but also the 'why' - the reasons that make this approach so potent and promising.

As you turn the pages of this book, my hope is that it will inspire you to embark on your own journey of discovery, whether you're a seasoned farmer, a home gardener, or a complete beginner. I hope it will empower you to experiment, to learn, and to witness the magic of this incredible world in your own garden or farm. And I hope that this book will contribute, even if in a small way, to the larger movement towards sustainable and environmentally friendly farming practices.

Introduction

Imagine, if you will, stepping into a world where the invisible forces of electricity dance with the tangible elements of biology. This is the realm of electroculture, a fascinating landscape where theory and practice intertwine in a mesmerizing ballet. This book, your guide, is an invitation to join this dance, to embark on a journey of discovery that will take you through the history, principles, and practical applications of electroculture.

Doesn't the very idea of it stir a sense of wonder within you? The thought of tracing the roots of electroculture back to its inception, back to the pioneering minds who first glimpsed the potential of electricity to enhance plant growth. This book will take you on that journey, illuminating the path with the light of scientific principles. It will reveal the intricate ways in which electricity interacts with biological systems and the environment, creating a symphony of growth and vitality.

But this book is not just a theoretical exploration. It's a call to action, a call to take the knowledge you've gained and apply it in a practical way. Can you picture it? Your garden, transformed into a thriving, vibrant ecosystem, powered by the invisible forces of electricity. To make this vision a reality, you need the right tools and materials. And this book will provide them.

Within these pages, you'll find detailed discussions on various tools used in electroculture, such as the Lightning Arrester, Spiral, Energy Tower, and Lakhovsky Ring. You'll also find step-by-step instructions for creating your own tools, empowering you to experiment with thi method in your own garden.

But the journey doesn't end there. This book will take you there, exploring emerging trends and innovations, as well as the challenges and opportunities that lie ahead. It acknowledges that while electroculture is a field brimming with potential, it remains relatively unknown and underutilized. But as awareness grows and research advances, we can expect to see it playing a more prominent role in sustainable agriculture.

In conclusion, this book is more than just a guide to electroculture. It's a journey of discovery and learning. It's a challenge that requires patience, curiosity, and a willingness to experiment. But for those who accept this challenge, the rewards can be immense. Not only does it offer a unique way to enhance plant growth and productivity, but it also provides a deeper understanding of the intricate interplay between electricity and life. So, whether you're a seasoned gardener looking to boost your yields, or a curious beginner eager to experiment with this fascinating field, this book has something for you. Are you ready to embark on this journey? Let's get started!

Chapter 1 - A Journey Through History: Electroculture Over the Years

As we set sail on this voyage of discovery, let's pause for a moment to marvel at the sheer beauty of exploration. The world we inhabit is a grand tapestry, woven with countless processes, each contributing its unique thread to the intricate design of life. One such thread, often unnoticed yet profoundly influential, is the role of electricity in agriculture - a practice known as electroculture.

In this chapter, we'll traverse the corridors of time, following the trail blazed by those audacious pioneers who dared to view the world through a different lens. These were the visionaries who recognized the potential in the invisible currents that pulse through our world. We'll delve into the early concepts and experiments, meet the notable figures who left their indelible mark, and explore the recent developments that have sculpted the landscape of electroculture.

But this journey is not merely a retrospective. It's also a telescope, allowing us to gaze into the future. It's about acknowledging the transformative potential of electroculture to revolutionize the way we cultivate our food and steward our environment. Can you picture it? A world where electricity and agriculture dance in harmony, creating a future that's more sustainable, more productive.
Are you ready to unravel the mysteries of this practice? If so, let's begin. For in the words of Lao Tzu, "The journey of a thousand miles begins with a single step"

Definition of Electroculture

What is electroculture, you might ask? To some, it may be a novel term, a fresh addition to their lexicon. Yet, it's a concept as old as time itself, recently resurfacing to the forefront of our modern consciousness.

At its heart, electroculture is the marriage of electricity and agriculture. It's the art of applying electricity or electric fields to the soil and plants, with the goal of enhancing their growth and yield. This practice, though steeped in centuries of history, has only recently begun to unveil its true potential. It's a captivating fusion of science and nature, a testament to our ability to harness the forces of the natural world for our benefit. But isn't it more than that?

Indeed, electroculture transcends the boundaries of a mere farming method. It's a philosophy, a lens through which we view the world. It's an acknowledgment of the intricate web of life, the subtle energies that course through the earth and all living beings. It's the understanding that by aligning ourselves with these energies, instead of resisting them, we can foster a more harmonious and productive relationship with nature.

The roots of this practice reach deep into the ancient world. Our ancestors, in their wisdom, recognized the power of the natural world, the vital energies that pulsed through the earth and the sky. They understood that by tapping into these energies, they could boost their crop growth and secure the prosperity of their communities.

As centuries rolled by, this ancient wisdom was overshadowed, replaced by more mechanistic approaches to agriculture. However, with the dawn of modern science, we are witnessing a renaissance of electroculture. We are beginning to comprehend the profound impact electricity, in its various forms, can have on plant growth and development. We are discovering that by introducing electric fields to our crops, we can not only enhance their growth and yield but also bolster their resistance to pests and diseases.

Yet, the magic doesn't stop at improving crop yields. It extends its nurturing touch to the very foundation of agriculture - the soil. Healthy soils are the bedrock of thriving crops, and electroculture has a pivotal role to play in preserving and enhancing soil health. By infusing the soil with electric fields, we can boost its fertility, improve its structure, and foster the growth of beneficial microorganisms.

In essence, this incredible method heralds a new paradigm in agriculture. It's a philosophy that weaves together threads of ancient wisdom and modern science. It's a perspective that acknowledges the power of the natural world and the life-giving energies that flow through the earth and the sky. More importantly, it's a beacon of hope for the future of our planet..

Early Concepts and Experiments

In the early days, when the concept of electroculture was just a seedling, it was curiosity and a thirst for understanding that watered its roots. Picture the first pioneers, their hands holding nothing more than a simple battery and a few wires, their minds filled with questions and wonder. These were the humble beginnings, the first brushstrokes on the canvas of a field that would eventually blend agriculture, electricity, and a profound reverence for nature into a beautiful masterpiece.

Can you feel the excitement of the mid-1800s? It was a time when the scientific discipline of electrophysiology was just beginning to spread its wings. Agricultural enthusiasts and scientists alike were in a race, a race to unlock the secrets of the earth, a race to revolutionize farming. And in this race, two currents of thought emerged.

One current, strong and swift, argued for the use of chemicals, fertilizers, and pesticides, promising a golden age of optimized plant growth. The other current, however, flowed in a different direction. It questioned the wisdom of relying on chemicals, voicing concerns about the potential harm to both the plants and the people who consumed them.

Guess which current swept the race? The proponents of chemicals emerged victorious, and thus began an era where chemicals became the lifeblood of agricultural operations. But the other current, the voice of dissent, was not silenced. It continued to flow, albeit in the shadows, a whisper of a thought that refused to die. This whisper would later echo through the halls of organic farming, a testament to the possibility of cultivating robust, healthy plants without the crutch of chemicals.

As the story of electroculture unfolded, it branched into two distinct paths. One path led to a system that harnessed human-generated electricity to nurture crops. Picture a network of wires, pulsating with electricity, creating an invisible dance of electric and electromagnetic fields that breathed life into the plants. The other path, however, chose to embrace the bounty of nature, harnessing the free energies that the Earth generously offered. This path, bathed in the gentle glow of natural energy, became the more popular choice. Yet, the first path was not abandoned. It still exists today, a testament to the diverse approaches within the realm of electroculture.

Why do you think the second approach gained more popularity? The answer lies in the fruits of its labor. The use of free natural energies has been observed to yield crops that outshine those cultivated with chemicals. These observations, collected over the years, have painted a compelling picture of the superiority of electroculture that harnesses natural energies.

But the canvas of this practice is vast and diverse. It encompasses techniques like magnetoculture, which taps into the Earth's natural magnetic fields to stimulate plant growth. The journey of it, from its inception in 1850 to the present day in 2023, is a testament to its validity. It is not a pseudoscience, nor is it a technique that lacks efficacy. Over 600 scientists, scattered across the globe, have witnessed the magic of electroculture. They have studied its effects, unraveled its mysteries, and offered a fresh perspective on the influence of magnetic, electric, and frequency fields on plant growth.

These pioneers have unearthed fascinating insights. They have discovered that frequencies affect not just plants and seeds, but also humans and animals. By manipulating these frequencies, we can enhance the bio-rhythm of plants, bolster their immune system, and make them more resilient to environmental stressors like drought or frost. We can boost their productivity and improve the quality of their yield, whether it's the lush green foliage, the vibrant flowers, or the succulent fruits.

These early experiments were not always successful, but they were always enlightening. They taught us about the complex interactions between electricity and living organisms

Notable Figures and Innovations

The tapestry of electroculture, rich and vibrant, is woven with the threads of numerous pioneers. Each thread, each contribution, is unique, adding depth and texture to the field. Can you feel the electric pulse of their collective efforts? It's this pulse that has shaped electroculture into the dynamic and promising discipline we see today.

Imagine the scene: Georges Lakhovsky, a French-Russian engineer, meticulously tending to geraniums, not with water and sunlight, but with the invisible currents of electricity. His invention, the Lakhovsky Ring, was a beacon of hope in the fight against plant cancer. Can you see the spark in his eyes as he realized the potential of manipulating the frequency of plant cells? His work was a significant leap forward, illuminating the intricate dance between electricity and plant growth.

Then there was Karl Selim Lemstrom, his mind buzzing with ideas as he penned "Utilization of Man-Generated Electrical Energies" in 1904. His work was a cornerstone in the edifice of electroculture, exploring the untapped potential of human-generated electricity. Can you sense the electricity in the air as he laid the groundwork for the techniques we use today?

Justin Etienne Christofleau, in his 1920 masterpiece "Cosmic Waves and Oscillating Circuits in Electroculture," painted a vivid picture of the relationship between electricity, electromagnetic fields, and plant growth. His theoretical contributions were like a compass, guiding us deeper into the labyrinth of electroculture.

Fast forward to the pivotal journal article "The Electroculture Of Crops," penned by Walter Stiles and Ingvar Jorgensen. Their research was like a lighthouse, casting a beam of light on the practical applications of electroculture in agriculture.

Picture Alexander Bain, his hands covered in soil as he embedded copper and zinc plates into the earth. His "Earth Battery," a humble yet revolutionary device, was a testament to the power of electricity to enhance plant growth.
In France, Father Paulin's work with an electro-vegetometer on a potato plant was a revelation. His observations were like a mirror, reflecting the transformation of plants into greener, healthier, and more productive versions of themselves. His work

was the seed that sprouted into Fernand Basty's first conference on electroculture.

In the modern age, the principles of electroculture found a champion in Bill Mollison, an Australian agronomist from the University of Tasmania. Known as the originator of Permaculture, Mollison's work was a symphony of electroculture principles.

In recent years, Yannick van Doorne, a Belgian engineer, has emerged as a torchbearer of the global electroculture movement. His work is like a ripple in a pond, spreading the innovative approach to a broader audience.

These luminaries, these pioneers, have each left their indelible mark on this field. Their contributions are like stepping stones, paving the way for future innovations and discoveries. As we stand on their shoulders, we look forward to the continued evolution of electroculture, to the dawn of a new era of sustainable and efficient agriculture.

Recent Developments

The world of electroculture is a dynamic landscape, ever-evolving and brimming with potential. Can you feel the pulse of its growth, the rhythm of its progress? It's a rhythm that has quickened over the years, a testament to the relentless pursuit of knowledge and innovation. Each year brings with it new techniques, new tools, and new insights. Farmers, those guardians of the earth, share their experiences, their triumphs, and their failures. They are the explorers charting new territories, the pioneers pushing the boundaries of what's possible.

Imagine the vast expanses of land in China, fields of rice, wheat, and corn stretching as far as the eye can see. Now, picture these fields bathed in the invisible glow of electroculture. This is not a distant dream, but a tangible reality. The trend is clear: this practice is expanding its reach, moving beyond the confines of a few hectares to encompass vast tracts of land. It's a revolution in the making, a revolution that promises to redefine the way we grow our food.

But the story of it doesn't end with the fields. It extends beneath the surface, to the very heart of the soil. Electricity, that invisible force, is being harnessed to enhance the transport of vital nutrients to the plants, to supercharge the process of photosynthesis, to breathe new life into the soil. The results are undeniable. The surge in electroculture adoption, particularly in Asia, speaks volumes about its effectiveness. It's a call to action, a call to invest in a future that promises healthier crops and sustainable farming practices.

Yet, it is not just for the farmers. It's for the gardeners, the plant enthusiasts, the green-thumbed individuals who find joy in the simple act of nurturing a plant. It's for those who wish to enhance the beauty of their ornamental plants, to infuse their gardens with the magic of electroculture. This trend underscores the versatility of electroculture, its ability to scale down and fit into the confines of a small garden or even an indoor plant pot.

The journey doesn't stop at the harvest. It continues, exploring new ways to enhance the nutritional value of the produce, to ensure that the benefits of electroculture extend to the dinner table. It's a journey fueled by creativity, by the desire to push the boundaries, to experiment with new systems and parameters. It's a journey that promises to usher in new ideas, alternative approaches, and cutting-edge tools.

But like any journey, the path of electroculture is not without its obstacles. Technical challenges loom large, threatening to stifle its growth. The potential of it, is vast, but the technical capabilities are still limited. Existing systems need to be refined, new systems need to be developed, and the benefits of large-scale electroculture need to be realized. The challenge is to strike a balance between efficiency and cost, to ensure that the benefits of electroculture are accessible to all.

As we stand on the cusp of a new era in agriculture, we look to the future with a sense of optimism and anticipation. The road ahead is fraught with challenges, but the promise of this practice is too great to ignore. As we delve deeper into the science behind electroculture in the next chapter, we remain hopeful. Hopeful that the power of electroculture will overcome the hurdles, hopeful that the future of agriculture will be shaped by this revolutionary approach.

Chapter 2 - Demystifying Electroculture: The Science Behind The Process

Let's embark on a voyage, a voyage that takes us deeper into this captivating universe. We're now at the epicenter, the pulsating heart where the magic happens - the science that breathes life into this process. This chapter is a treasure trove, a labyrinth of knowledge that unravels the complex tapestry of electroculture, breaking it down into threads of scientific principles that weave together to form a powerful tool in agriculture.

Electroculture, with its intricate web of complexities, may seem like a fortress, impenetrable to those without a background in physics or other scientific disciplines. But is it really so? Can the keys to this fortress be found in understanding the fundamental principles it stands on? Can these principles be harnessed to enhance plant growth, to breathe life into seeds and saplings? The answer is a resounding yes!

In this chapter, we're going to dive into the deep end. We'll explore the myriad ways electroculture can be applied in different agricultural practices. We'll share tips and tricks to amplify the effects of this practice. We'll delve into the role of electricity and electromagnetism in nature, those invisible forces that silently shape our world. We'll learn how to harness these natural forces, to channel them into stimulating plant growth.

Remember, we're not aiming to scale a mountain of scientific jargon or navigate a sea of complex concepts. Our goal is simpler, yet profound. We aim to gain a fundamental understanding of electroculture, to learn how to apply it in practical, real-world scenarios. So, are you ready?

Introduction to Electricity and Its Generation

Eletricity is an invisible force that powers our homes, fuels our vehicles, and even sparks life into our very bodies. But what is this enigmatic entity we call electricity? How does it come into being?

At its heart, electricity is a form of energy born from the existence of charged particles, such as electrons or protons. These particles can either be static, forming an accumulation of charge, or dynamic, creating a current. This energy, as elusive as it may be, can be harnessed and put to work, powering a light bulb or driving a motor.

But how do we generate this energy? The answer lies in a myriad of methods, each with its unique processes and principles. The most common method is through the use of generators. Picture a wire, a simple conductor, dancing through a magnetic field. This dance induces a flow of electrical current, transforming mechanical energy into electrical energy.

Then there are batteries, silent powerhouses of chemical reactions. Within a battery, a chemical reaction sparks a flow of electrons from one terminal to another, birthing an electric current. This process continues until the chemicals are spent, and the battery, now 'dead', needs to be recharged or replaced.

Consider also the magic of solar panels. They generate electricity through a process known as the photovoltaic effect. Imagine sunlight, the life-giving rays of our star, hitting the solar panel. This excites the electrons in the panel's semiconductor material, stirring them into motion and creating an electric current.

In the realm of electroculture, electricity can be generated through natural means, such as atmospheric electricity or telluric currents. Atmospheric electricity is a fascinating study of electrical charges in the Earth's atmosphere. Picture the Earth, the atmosphere, and the ionosphere engaged in a grand dance, the global atmospheric electrical circuit, where charges move and mingle. Telluric currents, on the other hand, are electric currents that flow underground or through the sea. These currents, born from both natural causes and human activity, interact in a complex pattern, like dancers in a ballet.

The generation and application of electricity in electroculture can be a complex process, a delicate balancing act of various factors. The type of current, the intensity of electron flow, the voltage, and even the type of electrodes used can all influence the effectiveness of the electroculture system. It's a puzzle that requires a comprehensive understanding of these factors and how they interact.

Studies have also suggested that we can harness energy directly from the Earth. Our planet, a giant battery, carries a natural electrical charge. By tapping into this energy source, we can utilize the Earth's inherent electrical properties for various purposes within electroculture. This revelation has opened up new avenues for sustainable energy generation and broadened the potential applications of electroculture across diverse fields.

As we delve deeper into this book, we will explore in detail how to exploit this type of energy. Whether you are a novice just dipping your toes into this world, or an experienced practitioner, this book aims to equip you with the knowledge and tools to harness the power of the Earth's energy in your electroculture practices.

Role of Electricity In Nature

Electricity is a fundamental part of nature, playing a crucial role in various natural processes. From the spectacular displays of thunderstorms and auroras to the subtle electrical signals within living organisms, electricity is an integral part of the world around us.

Have you ever paused to marvel at the raw power of a thunderstorm? The way the sky darkens, the air crackles with anticipation, and then, in a brilliant flash, electricity arcs across the sky. This spectacle is one of nature's most dramatic displays of electricity.

Picture this: within the tumultuous heart of a cloud, air currents surge upward and downward, causing a separation of charges. The top of the cloud bristles with positive charges, while the bottom hums with negative ones. When the tension between the cloud and the ground becomes unbearable, the sky splits open in a dazzling discharge of electricity - lightning. But the spectacle doesn't end there. The rain that follows is enriched with nitrogen, a vital nutrient that gives plants a significant boost. Isn't it fascinating how a storm's fury can nurture life?

Now, let's journey from the tempestuous drama of a thunderstorm to the serene beauty of the Aurora Borealis, or the Northern Lights. This is another stunning display of nature's electricity. Charged particles from the sun collide with the Earth's magnetic field, and the energy from this cosmic dance is transferred to the atoms and molecules in our atmosphere, causing them to glow in various colors. This process is a testament to the far-reaching effects of electricity, extending even to the interaction between our planet and the sun.

Did you know that the Aurora Borealis is thought to be responsible for the lush vegetation in the Arctic? According to experimental work by Professor Selim Lemström in 1904, the electrical fields from these celestial lights could have a beneficial impact on plant life. This insight into the Aurora Borealis's influence on plant growth provides a fascinating glimpse into the potential of electroculture.

But the role of electricity in nature isn't limited to these grand displays. It's also at work in the subtle, unseen processes within living organisms. Consider our bodies, for instance. We use electrical signals to transmit information between cells, allowing us to move, think, and feel. Similarly, plants use electrical signals for various processes, such as responding to environmental stimuli and coordinating their growth.

In the realm of electroculture, understanding the role of electricity in nature is paramount. Phenomena such as atmospheric electricity and telluric currents can be harnessed to enhance plant growth. For instance, it has been observed that plants often exhibit enhanced growth following a thunderstorm. Similarly, the presence of telluric currents in the soil can influence the uptake of nutrients by plant roots, affecting their growth and productivity.

This understanding of natural electrical phenomena and their impact on plant growth is the cornerstone of electroculture. By harnessing these forces, we can cultivate healthier, more robust plants, even in small gardens or indoor plant setups. This makes electroculture an exciting option for both beginners and experienced gardeners.

Explanation of The Earth Battery Experiment

The year is 1841, a time when the mysteries of electricity were just beginning to be unraveled. Amidst this era of discovery, a Scottish inventor named Alexander Bain embarked on an experiment that was as audacious as it was simple. His idea? To create an 'earth battery', a device that could harness the electrical power of the earth itself. Can you imagine the thrill of such a prospect?

Bain's experiment involved inserting plates of zinc and copper into the fertile earth, connected by a wire that danced in the breeze above the ground. Then came the truly remarkable part. When plants were nestled in the soil between these plates, they began to flourish, growing at a rate and to a size that was noticeably greater than their counterparts in regular soil.

This Earth Battery Experiment, as it came to be known, was a watershed moment in the evolution of electroculture. It provided the first tangible proof that the earth's inherent electrical properties could be harnessed to enhance plant growth. This revelation ignited a spark that would inspire a wave of further research and experimentation.

Bain's experiment was grounded in the understanding that the earth itself is a symphony of electrical activity. By inserting metal plates into the ground and connecting them with a wire, Bain was able to generate a small electrical current. This current, when introduced to the soil, seemed to breathe life into the plants, enhancing their growth and development.

But the Earth Battery Experiment was not a solitary event, destined to be forgotten in the annals of history. On the contrary, it was replicated by numerous researchers over the years, each time yielding positive results. For instance, researcher George Hull created his own version of an earth battery "in a box" and observed increases in plant growth of 20 to 40 percent. He attributed this growth spurt to the current flowing through the soil, a testament to the power of Bain's original experiment.

The Earth Battery Experiment stands as a beacon, illuminating the potential of electroculture. This concept, as relevant today as it was in the mid-1800s, forms the cornerstone of our exploration into the science behind electroculture. As we delve deeper into this exciting field, we will explore how these principles can be applied in practical scenarios, from nurturing a small indoor plant to managing large-scale agricultural operations. The Earth Battery Experiment is just the beginning.

Chapter 3 - Soil: The Foundation of Agriculture

Imagine standing at the precipice of a vast, unexplored world, a world teeming with life, complexity, and untold potential. This world is not in the far reaches of space, but right beneath our feet. It's the soil, the unsung hero of our planet, the silent nurturer of life. Often overlooked, the soil is a dynamic ecosystem, a bustling metropolis of minerals, organic matter, water, air, and countless organisms. Can you feel its pulse? Can you sense its vitality?

In the grand theatre of life, soil is not a mere spectator. It's an active participant, a vital character that shapes the narrative of plant growth. It's a living entity, a complex tapestry woven with threads of life, each thread playing a crucial role in nurturing plant life.

Now, imagine introducing electricity into this vibrant tableau. This is where the magic of electroculture begins to unfold. In the world of electroculture, soil takes center stage. It becomes the conduit, the life-giving river that carries the electrical currents, influencing the growth and development of plants.

But what makes a soil 'good'? What gives it the power to nurture life, to cradle the seeds of growth? As we delve into the intricate world of soil science, we'll explore these questions. We'll examine the characteristics of fertile soil, unravel the secrets that make it conducive for plant growth, and discover how these characteristics can be enhanced through electroculture.

The health of the soil is the heartbeat of electroculture. A healthy soil is a thriving ecosystem, a balanced symphony of elements that work in harmony to support plant life. But how can we ensure the health of our soil? How can we maintain its vitality and enhance its potential? As we journey deeper into this chapter, we'll explore these questions. We'll discover the steps we can take to maintain and improve soil health, and in doing so, enhance the success of our electroculture practices.

As we embark on this journey, you'll gain a newfound appreciation for the soil beneath our feet.

*This will be further explored in chapter 7

Characteristics of Fertile Soil

The soil is a vibrant ecosystem teeming with life. It's a symphony of organic matter, minerals, gases, liquids, and a myriad of organisms that together form the foundation of life on Earth. In the realm of electroculture, this humble soil takes center stage, its characteristics forming the cornerstone of effective practices.

Picture the soil, not as a monotonous mass, but as a mosaic of diverse elements. It's a blend of air, water, mineral particles, organic matter, and a plethora of organisms. Imagine the tiny spaces between soil particles, the pore spaces, half-filled with air and half with water. Can you see how these spaces serve as conduits for the movement of water, nutrients, and air, creating the perfect conditions for plant roots to breathe and absorb nutrients?

The texture of the soil, a tapestry woven from clay, silt, and sand, is a critical determinant of its characteristics. Each type of soil, whether it's loamy, sandy, or clayey, tells a unique story. A story of its ability to retain water and nutrients, its aeration, and its compatibility with different types of plants.

Now, let's delve deeper into the soil, into the realm of organic matter, often referred to as humus. This is the lifeblood of the soil, holding soil particles together, storing nutrients, and nourishing soil organisms. It's a testament to the cycle of life, derived from the decomposition of plant and animal material, enriching the soil with nutrients, and enhancing its structure and water-holding capacity.

The soil is not just a physical entity, but a living, breathing ecosystem. Microorganisms, like bacteria and fungi, are the unsung heroes of this ecosystem. They break down organic matter, releasing nutrients into the soil, and creating compounds that bind soil particles together. They are the silent custodians of nutrient cycling and soil fertility.

The fertility of the soil is also a dance between chemistry and biology. The pH of the soil, a measure of its acidity or alkalinity, is a key player. It influences the availability of nutrients in the soil, and different plants sway to the rhythm of different pH levels.

In the world of electroculture, the soil dons another hat, that of an electrical conductor. The soil's ability to conduct electricity, its electrical conductivity, can significantly influence the effectiveness of electroculture practices. Factors like the soil's moisture content, the concentration of dissolved electrolytes, and its composition, all play a part in this electrical symphony.

Understanding the soil is akin to understanding the language of the land. In electroculture, this understanding is vital. For instance, the electrical stimulation of soil can induce the movement of water and nutrients. Clay-rich soils, with their high ionic mobility, respond beautifully to this stimulation, enhancing plant growth not just through nutrient availability, but also through increased nutrient mobility. Sandy soils, on the other hand, do not dance as well to this electrical tune, their nutrient mobility is not as high, and hence, the impact of electrical stimulation is not as profound.

So, the next time you walk on the soil, remember, you are walking on a dynamic, living entity. An entity that holds the key to effective electroculture practices, an entity that is more than just the dirt beneath our feet.

Importance of Soil Health In Electroculture

In the world of agriculture, the health of the soil is paramount. It's the bedrock of any successful cultivation, the canvas on which the masterpiece of a bountiful harvest is painted. The importance of soil health in electroculture is no different. It's like a symphony where each element plays a crucial role in creating a harmonious outcome.

The soil is a living entity, a complex ecosystem that thrives on balance and care. It's a symphony where each element, each microorganism, each nutrient, plays a crucial note. The melody they create together is the song of fertility, the music of life. But what happens when we introduce a new instrument into this symphony, the instrument of electricity?

In the world of electroculture, the texture of the soil takes center stage. Medium-textured soils, a harmonious blend of sand, clay, silt, and humus, are the ideal dance partners for electricity. They have a structure that allows for the smooth flow of electricity, a rhythm that syncs perfectly with the pulse of electroculture. But the dance of fertility is not just about the physical steps; it's also about the emotional connection, the chemistry between the partners.

The organic component of the soil, the decomposing plant and animal matter, the bacteria, the fungi, they are the soul of the soil. They break down organic matter into a buffet of nutrients, a feast for the plants. They contribute to the structure of the soil, making it more porous, more capable of holding onto water and nutrients. They are the unseen heroes, the backstage crew that sets the stage for the spectacle of growth.

In the grand theater of electroculture, the health of the soil is the lead actor. A healthy soil can support a cast of robust, resilient plants, plants that can stand tall against the onslaught of pests and diseases. It can provide the nutrients, the sustenance that the plants need to thrive. But the magic is not just about leveraging the power of a healthy soil; it's also about enhancing it..

The application of electricity in electroculture is like a magic wand, transforming the soil, making it more aerated, more capable of retaining water and nutrients. It's like a conductor, orchestrating the symphony of microorganisms, enhancing the breakdown of organic matter, and the release of nutrients. But like any powerful magic, it needs to be wielded with care.

The power of the system must be balanced, like a tightrope walker maintaining their poise. Too much power, and the crop could be overwhelmed, drowned in a flood of nutrients. Understanding the health of the soil, adjusting the electroculture system accordingly, is the key to maintaining this balance.

In the end, the health of the soil is the foundation on which the edifice of successful cultivation is built. By understanding the characteristics of fertile soil and the role of electricity in enhancing soil health, we can unlock the full potential of electroculture.we grow our food.

Chapter 4: Benefits Of Electroculture

As we delve deeper into the world of electroculture, we begin to uncover the numerous benefits that this innovative approach to agriculture has to offer. The fusion of electricity and agriculture is not just a scientific curiosity; it's a practical tool with the potential to revolutionize the way we grow our food.

In this chapter, we're about to embark on an exploration, a journey that will reveal the myriad benefits of electroculture. From accelerated plant growth and robust health to bountiful crop yields, the advantages are as diverse as they are impressive. But how does it work? How does this fusion of electricity and agriculture yield such remarkable results? Let's delve into the science, unravel the intricacies, and gain a comprehensive understanding of how electroculture can elevate agricultural practices and contribute to sustainable farming.

Electroculture, you'll discover, is not merely a cultivation method. It's a holistic approach to agriculture that takes into account the vitality of the soil, the well-being of the plants, and the sustainability of farming practices. It's about harnessing the power of electricity to amplify the natural processes of growth and development, leading to crops that are not just healthier, but also more resilient.

In the sections that follow, we'll delve deeper into these benefits. We'll explore how electroculture can rejuvenate soil health, enhance seed germination, boost crop growth and yield, and even contribute to integrated pest management.

But as we uncover the benefits of this practice, we must also acknowledge the challenges and considerations that come with it. Electroculture is not a one-size-fits-all solution. It demands a nuanced understanding of the interplay between electricity and biological systems, and it must be tailored to the specific needs and conditions of each garden/farm.

Yet, armed with the right knowledge and tools, electroculture can be a formidable ally in our quest for sustainable, resilient agriculture. So, are you ready? Let's embark on this journey of discovery, uncover the myriad benefits of this field, and explore how it can revolutionize the way we cultivate our food.

Faster Plant Growth

The world of electroculture is a fascinating one, filled with potential and promise. One of the most significant benefits it offers is the acceleration of plant growth. This isn't a mere hypothesis; it's a fact backed by numerous studies and practical applications.

The principle behind this accelerated growth lies in the interaction between the electric field and the biological processes within the plant. When an electric field is applied to the soil, it stimulates the movement of ions in the soil and within the plant itself. This enhanced ionic transport can lead to increased nutrient uptake and more efficient metabolic processes, resulting in faster growth rates.

What does this mean for us, for the farmers who toil in the fields, for the consumers who rely on their harvest? It means crops that are ready for harvest sooner, a boon for farmers, especially those grappling with short growing seasons or capricious weather. Imagine being able to outpace the frost, to bring in the harvest before the winter winds begin to howl. This is the power of electroculture.

Imagine fields that are never fallow, where the harvest of one crop is quickly followed by the planting of the next. This is the potential of electroculture, a world where multiple planting cycles within a single season are not just a dream, but a reality.

Yet, like any magic, electroculture requires a careful hand. The dance of the ions, the strength of the electric field, the rhythm of the soil, all must be in harmony. Each crop, each field, each electric field is unique, and finding the perfect balance requires experimentation, observation, and patience.

Now, let's step back and look at the bigger picture. Our world is growing, with the global population projected to reach 9.7 billion by 2050. The question looms: how will we feed everyone? Electroculture, with its ability to accelerate plant growth and increase crop yields, could be a crucial part of the answer.

And what about sustainability, the call to nurture our planet even as we draw from its bounty? By speeding up the growth of plants, we can reduce the resources needed for each crop, creating a more efficient, more sustainable agricultural system.

In the final analysis, the world of electroculture is a world of possibilities. It offers a path to increased productivity, to enhanced sustainability, to a future where the bounty of the fields keeps pace with the needs of the world. It's a world where the growth of a plant is not just a natural process, but a testament to the power of human ingenuity.

Healthier Plants And Increased Flavor

Can you imagine biting into a fruit that bursts with flavor, its sweetness and tanginess perfectly balanced, its texture crisp and satisfying? This is not a mere fantasy, but a tangible reality made possible by the fascinating world of electroculture. This journey into the realm of electroculture is not just about healthier plants, but also about a sensory experience that tantalizes our taste buds.

What makes plants healthier in the world of electroculture? The answer lies in the invisible dance of nutrients and metabolic processes, all orchestrated by the silent conductor - the electric field. Imagine the soil teeming with nutrients, each particle charged with the potential to nourish a plant. Now, picture an electric field, its invisible lines of force stimulating the movement of these nutrient-rich particles. The result? Plants that are not just healthier, but also more resilient. Plants that can stand tall against the onslaught of pests and diseases, reducing our reliance on chemical pesticides and nudging us towards a more sustainable and organic farming system.

Have you ever wondered why some fruits burst with flavor while others taste bland? The secret lies in the concentration of sugars, acids, and other flavor compounds, which are enhanced by the nutrient uptake in electroculture. Imagine biting into a tomato that's been nurtured with electroculture. Can you taste the sweetness, the tanginess, the rich umami flavor that sets it apart from its conventionally grown counterparts?

The science behind this is as fascinating as the flavors it enhances. Picture the electric field as a maestro, directing the movement of ions in the soil and within the plant itself. This symphony of ionic transport leads to increased nutrient uptake and more efficient metabolic processes. The result? Fruits and vegetables that are not just a feast for the taste buds, but also a powerhouse of nutrients.

Resilience is another virtue that electroculture bestows upon plants. Picture a plant standing tall and robust, its leaves a vibrant green, its fruits plump and juicy. This plant is not just surviving, but thriving, equipped with the strength to withstand pests and diseases. This resilience reduces our reliance on chemical pesticides, contributing to a more sustainable and organic farming system. Isn't this what we need in a world that's increasingly conscious of the environmental impact of farming practices?

And let's not forget about the post-harvest benefits. Imagine a world where fresh produce stays fresh for longer, where the crunch of a bell pepper or the juiciness of a peach doesn't fade away in a few days. Healthier plants, nurtured by electroculture, yield produce that can be stored for longer after harvest. This not only reduces waste but also ensures that fresh produce graces our tables for longer.

In the grand tapestry of electroculture, the threads of healthier plants, enhanced flavors, and increased sustainability are intricately woven together. It's a solution to some of the most pressing challenges facing the agricultural industry today. It's about enhancing the nutritional quality and flavor of our food, about reducing the environmental impact of farming, and about creating a future where agriculture and nature exist in harmony. Isn't that a future worth striving for?

Higher Crop Yields

In the grand theater of agriculture, success is often gauged by the bounty of the harvest. The more plentiful the yield, the louder the applause. But what if there was a way to amplify this applause, to make the yield even more abundant? This is where the magic of electroculture takes center stage, casting a spell that significantly boosts crop yields. But how does this magic work?

At the core of electroculture's success in enhancing crop yields lies the principle of plant health and growth. Imagine a plant, its leaves reaching out towards the sun, its roots delving deep into the nourishing soil. Now imagine this plant on an energy boost, its growth accelerated, its health enhanced. This is what electroculture does. It supercharges the plants, and in doing so, it supercharges the yield. But this fantastic practice doesn't stop at healthier plants. It goes a step further, stimulating physiological responses in plants that turbocharge their productivity.

Picture the soil as a bustling marketplace, teeming with nutrients. Now picture an electric field, acting as a catalyst, enhancing the movement of ions in this marketplace. This is one of the key factors that contribute to higher yields in electroculture. The electric field boosts nutrient uptake, allowing the plants to feast on a richer spread of nutrients. And when plants feast on more nutrients, they grow larger and produce more.

But the electroculture doesn't stop at nutrient uptake. Research has shown that electric fields can stimulate the production of these hormones, leading to faster and more vigorous growth. Imagine a plant on a growth spurt, its leaves unfurling faster, its roots spreading wider. This is the result of the stimulation of plant growth hormones, leading to larger plants and consequently, higher yields.

Electroculture is not just about boosting productivity; it's also about efficiency. It's about doing more with less. For instance, plants nurtured under the care of this method require less water. They're like prudent consumers, making the most of every drop. This not only conserves water but also boosts productivity, leading to higher yields.

However, too much electricity can harm the plants, while too little may not yield the desired results. It's a delicate dance, a dance that requires understanding and balance.

In conclusion, the promise of higher crop yields is one of the most compelling aspects of electroculture. It offers a practical and sustainable solution to one of the biggest challenges facing the agricultural industry today: how to produce more with less.

Other Benefits Of Electroculture

The realm of electroculture, as we've explored so far, is a treasure trove of benefits that extend their generous hands beyond merely enhancing plant growth and yield. But isn't there more to this fascinating field? Indeed, there is. Electroculture, in its versatile glory, offers a plethora of additional advantages that make it a formidable tool in the agricultural world.

Consider the battle against pests, a constant struggle for farmers worldwide. Here, electroculture emerges as an unexpected ally. Plants nurtured under the watchful gaze of electric fields seem to lose their allure to pests. Could it be a shift in the plant's metabolic rhythm or the birth of certain compounds that pests find repugnant? Whatever the cause, the result is a potential reduction in the use of chemical pesticides, paving the way for a greener, more sustainable farming system. Isn't that a future we all yearn for?

And what about the silent heroes of agriculture, the pollinators? The invisible dance of electric fields appears to charm these pollinators, drawing them towards the crops and boosting pollination. Imagine the joy of farmers witnessing more bountiful and consistent yields, especially for crops that rely heavily on these tireless insects.

But the gifts of electroculture don't stop at the boundaries of the field. It also offers a shield against the unpredictable whims of weather. The accelerated growth rates could allow farmers to gather their harvests earlier, potentially outwitting early frosts or other adverse weather conditions. And the plants? They stand tall and robust, their resilience enhanced, ready to brave the trials of drought.

The soil, the very cradle of agriculture, too, reaps the benefits of electroculture. The electric fields seem to breathe life into beneficial soil microorganisms, leading to improved soil structure and fertility.

And if the soil is tainted with pollutants? The electric fields could stimulate the activity of certain microorganisms capable of breaking down these pollutants, offering a potential solution for soil remediation that is both cost-effective and kind to our planet.

Lastly, let's turn our gaze to one of the most precious resources on Earth - water. The plants nurtured by electroculture, in their robust health, require less water. The electric fields could also enhance the soil's ability to hold water. Could this lead to more efficient use of water, a boon in regions where water is a scarce commodity?

In conclusion, the additional benefits of electroculture paint a picture of a versatile and powerful tool that could revolutionize the agricultural industry. From managing pests to rejuvenating soils, this method offers practical, sustainable solutions to a myriad of challenges. Isn't it time we embraced the full potential of this remarkable field?

Chapter 5 - The Heart of the Matter: Understanding Electroculture Systems

As we delve deeper into this world we arrive at the heart of the matter: understanding electroculture systems. This chapter will guide you through the intricate workings of these systems, shedding light on their structure, operation, and the science that underpins them.

Electroculture systems are not merely a jumble of wires and electrodes. No, they are far more than that. They are masterpieces of design, meticulously crafted to harness the raw power of electricity and channel it into a life-giving force for plants. Understanding these systems is akin to learning a new language, a language that allows us to converse with nature in ways we never thought possible.

Now, in this vast landscape, two paths stretch out before us. One path leads us to the realm of man-made electric energy, while the other beckons us towards the bounty of free, natural energies. Each path, with its unique allure and challenges, offers a different journey. Which path will you choose? Your choice may depend on your goals, the resources at your disposal, and the principles that guide your journey.

In this chapter, we will venture down the more complex path, exploring the labyrinth of electroculture systems that exploit energy generated by man. These systems, while demanding a deeper understanding and meticulous setup, promise a greater degree of control and potential rewards. We will delve into the basics of these systems, explore the different types that exist, and guide you through the process of setting up and maintaining your own.

However, we understand that for beginners, the simpler path may be more appealing. Systems like Lakhovsky rings, spirals, and pyramids offer a gentler introduction to this world. While they may not offer the same level of control as their more complex counterparts, they still hold significant benefits and provide a perfect playground for your first foray into electroculture.

In the following chapter, I will guide you through the process of setting up these simpler systems. I will provide step-by-step instructions and practical tips to help you embark on your journey in a way that is accessible and manageable, even for beginners.

By the end of these two chapters, you will have traversed the vast landscape of electroculture systems. You will be armed with the knowledge and skills to harness the power of electricity for plant growth.

The Basic of Electroculture Systems

Electroculture systems are a fascinating blend of science and nature, harnessing the power of electricity and natural energies to stimulate and enhance plant growth.
At the heart of an electroculture system is a principle as old as time itself - the use of electricity to stimulate growth. But how does this happen? Picture a sphere of electricity enveloping the plants, a sphere that pulses with life, influencing the growth and development of the plants within. This sphere can be conjured in various ways - through electrodes that plunge into the soil like roots, through electric fields that wrap around the plants like a mother's embrace, or through pulsed electromagnetic fields that rhythmically pulse, like the beating of a heart.

Have you ever thought about the Earth as a battery? With its positive and negative terminals, a battery is a microcosm of the Earth and its atmosphere. Together, they form a natural electrical circuit, a global electric circuit that is an integral part of our planet's electromagnetic environment. It's within this environment that all life, including plants, have evolved. In the world of electroculture, this natural electrical potential is not just observed, it's harnessed, tamed, and directed to serve the growth of plants.

The electricity used is as diverse as the plants it nurtures. Its type, intensity, frequency, and duration of application are all variables in the equation of plant growth. Some plants might thrive under the gentle touch of a constant, low-intensity electric field, while others might bloom under the sporadic bursts of high-intensity pulses.

The design of an electroculture system is a puzzle with many pieces. Factors such as the type of soil, the climate, the species of plants, and their growth stage are all pieces that need to fit together perfectly to create an effective system. It's a delicate balancing act, a dance of variables that need to be in perfect harmony.

In essence, the art of electroculture is about understanding the intricate relationship between electricity and plant growth. It's about leveraging this knowledge to create an optimal growing environment for plants. It's about working in harmony with nature, not against it, to achieve better results in agriculture. It's about harnessing the power of electricity to whisper to the plants, to guide them, to help them grow. It's about becoming an alchemist of the agricultural realm.

Types of electroculture systems

Electroculture systems come in a variety of forms, each with its unique characteristics and benefits. The choice of system largely depends on the specific needs of the plants, the conditions of the environment, and the goals of the gardener or farmer. Here, we will explore some of the main types of electroculture systems.

Direct Current (DC) Systems: These systems use a direct current of electricity, typically supplied by a battery or a solar panel. The electricity is applied to the soil through electrodes, creating an electric field that can stimulate plant growth. DC systems are relatively simple to set up and can be used in both small-scale and large-scale farming.

Alternating Current (AC) Systems: AC systems use an alternating current of electricity, which changes direction periodically. This type of system can create a more dynamic electrical environment, which some studies suggest may be more beneficial for plant growth. However, AC systems are generally more complex and require more equipment than DC systems.

Pulsed Electromagnetic Field (PEMF) Systems: PEMF systems use pulsed electromagnetic fields to stimulate plant growth. These systems can be highly effective, but they are also more complex and require more specialized equipment. PEMF systems are typically used in more advanced or experimental forms.

Natural Electroculture Systems: These systems aim to harness the natural electrical potential of the Earth and the atmosphere (see the next chapter)

Hybrid Systems: Some systems combine elements of the above types to create a hybrid approach. For example, a system might use a DC current to create a basic electric field, and then supplement this with PEMF pulses to provide additional stimulation.

Each of these systems has its strengths and weaknesses, and the choice of system will depend on a variety of factors. In the following sections, we will delve deeper into these systems, exploring how to set them up and how to use them effectively. We will also discuss the scientific theories behind these systems, providing a deeper understanding of how and why they work.

DIRECT CURRENT (DC) SYSTEMS IN ELECTROCULTURE:

Imagine a river, its waters flowing steadily in one direction, carving a path through the landscape. This is the essence of Direct Current (DC) systems, one of the most prevalent types of electroculture systems. They function on a principle as simple and as profound as a river's flow: a ceaseless stream of electric charge moving in a single direction. The source of this flow? A humble battery or a solar panel, making DC systems a beacon of simplicity in the world of electroculture.

Now, picture this electric current coursing through the veins of the soil, delivered by electrodes crafted from conductive materials like copper or iron. These electrodes, plunged into the heart of the earth, become conduits for the electric current, creating an invisible arena of electric fields around the plants. Can you imagine the impact? This electric field becomes a catalyst, spurring plant growth by enhancing nutrient uptake, fueling cell division, and stimulating the growth of roots.

The beauty of DC systems lies in their simplicity. They are devoid of the need for intricate equipment or advanced technical knowledge, making them a canvas that can be painted by the hands of a novice gardener or a seasoned farmer. Their simplicity, however, does not compromise their versatility. They are equally at home in a small-scale home garden or a sprawling commercial farm.

But, like a river that changes its course in response to the landscape, the effectiveness of a DC system is not a constant. It ebbs and flows, influenced by a myriad of factors. The intensity of the electric current, the type of soil, the species of plants, and the specific environmental conditions - all these elements shape the effectiveness of a DC system. Isn't it fascinating how some plants thrive under the gentle caress of a low-intensity electric field, while others bloom under the more forceful touch of a higher intensity?

The versatility of DC systems extends beyond their application. They can be tailored to meet specific needs. The electrodes, for instance, can be strategically placed to target specific areas of the soil. The intensity of the electric current can be fine-tuned to provide the optimal level of stimulation for the plants.

In essence, DC systems are a testament to the power of simplicity and flexibility in the realm of electroculture. They offer a practical and adaptable approach to harnessing the power of electricity to enhance plant growth. Whether you're a gardener seeking to invigorate your plants or a farmer exploring sustainable farming practices, DC systems are a valuable ally in your journey through the captivating world of electroculture.

PULSED ELECTROMAGNETIC FIELD (PEMF) SYSTEMS IN ELECTROCULTURE:

These systems, like a master sculptor, shape and mold electromagnetic fields into pulses, using them as a catalyst to stimulate plant growth. Isn't it fascinating to think that something as intangible as electromagnetic pulses can have such a tangible impact on the biological processes of plant growth and development?

PEMF systems operate by generating short, controlled bursts of electromagnetic energy. Imagine being able to fine-tune the frequency, intensity, and duration of these pulses, creating a tailored electrical environment for your plants. This adaptability makes PEMF systems a versatile tool, capable of catering to a diverse range of plant species and growth conditions.

Research has painted a promising picture of the potential benefits of PEMF systems. They have been shown to stimulate cell division, enhance nutrient uptake, and bolster resistance to disease. Can you imagine the result? Healthier plants and higher crop yields, a dream come true for both home gardeners and commercial farmers.

However, as with all advanced technologies, PEMF systems come with their own set of challenges. They require more specialized equipment, including a PEMF generator, the heart of the system that pumps out the required electromagnetic pulses. These generators can be quite an investment and require a certain level of technical prowess to operate effectively.

Moreover, the science behind PEMF systems, while promising, is still shrouded in mystery. The exact mechanisms by which they influence plant growth are yet to be fully unraveled. This means that there is still a vast ocean of knowledge waiting to be explored, making ongoing research and experimentation crucial.

In conclusion, PEMF systems, while offering a high degree of control and potential benefits, also demand a significant investment in terms of equipment and knowledge. They are like a challenging mountain peak, best suited for those who are experienced climbers in the realm of electroculture, or those who are willing to invest the time and resources required to conquer this peak.

NATURAL ELECTROCULTURE SYSTEMS IN ELECTROCULTURE

The realm of natural electroculture systems is a testament to the awe-inspiring power of our planet. These systems seek to tap into the Earth's inherent electrical potential, a silent force that pulses through the soil beneath our feet and the air we breathe. Can you imagine? Our planet and its atmosphere, a vast, natural electrical circuit, waiting to be harnessed to invigorate plant growth.

The secret to mastering natural electroculture systems lies in the art of understanding and aligning with the environment's inherent electrical properties. Consider the Earth's magnetic field, the electric charge suspended in the atmosphere, or the electric essence of soil and water. Each of these elements holds the potential to influence plant growth, to breathe life into seeds and saplings.

Imagine aligning your planting rows with the Earth's magnetic field, a subtle adjustment that could enhance the electrical environment cradling your plants, coaxing them into growth. Or consider the use of naturally occurring materials, like specific rocks or minerals, known to enhance the soil's electrical properties. It's like giving your plants a natural, gentle boost, a secret whisper that encourages them to thrive.

The beauty of natural systems is their simplicity. They don't demand complex equipment or external power sources. Instead, they ask for understanding, for a keen awareness of the environment, and a grasp of the electrical properties that can coax plants into growth. (Chapter 6)

What makes natural systems truly remarkable is their harmony with nature. They don't seek to force growth through artificial means, but rather to enhance the natural growth processes of plants. The result? Healthier plants, nurtured by the Earth's own energy, and a more sustainable approach to agriculture.

In essence, natural electroculture systems offer a unique, sustainable approach to enhancing plant growth. They demand an understanding of the environment and a willingness to work in harmony with nature. For those who yearn for a more natural, sustainable approach to agriculture, these systems are not just tools, but allies.

HYBRID ELECTROCULTURE SYSTEMS IN ELECTROCULTURE:

The aim of a hybrid system is akin to creating a perfect recipe. It seeks to harness the strengths of each system while minimizing their weaknesses. Consider this: a hybrid system might employ a DC system to provide a consistent base level of electrical stimulation, supplemented by periodic pulses of electromagnetic energy from a PEMF system. Doesn't this combination sound like it could offer a more balanced and effective form of stimulation than any single system alone?

The beauty of hybrid systems lies in their adaptability. They can be fine-tuned to cater to the specific needs of the plants and the environment. Imagine being able to adjust the frequency and intensity of the PEMF pulses to align with the growth cycle of the plants, or tweaking the DC current to match the conductivity of the soil. This level of control paves the way for customization and optimization.

However, with great power comes great responsibility. Hybrid systems, while potent, are also the most complex type of electroculture system. They demand a deep understanding of the principles of electricity and electromagnetism, coupled with a solid grasp of plant biology and soil science. They also call for more equipment and resources than other types of systems.

Setting up and managing a hybrid system is akin to conducting an orchestra. It requires meticulous planning and coordination to ensure that all the different elements of the system are working in harmony. This can be a daunting task, especially for those who are new to the world of electroculture.

In conclusion, hybrid systems are like a master key, offering the potential for a highly effective and versatile approach to electroculture. They amalgamate the strengths of different types of systems to provide a comprehensive solution. However, they also demand a significant investment in terms of knowledge, equipment, and resources. Therefore, they may be best suited to those who are ready to dive deeper into the realm of electroculture, those who are not afraid to experiment and explore.

Setting Up an Electroculture System

Venturing into the realm of electroculture can be an exhilarating expedition, a voyage into the captivating crossroads of electricity and biology. The process may seem intimidating at first glance, but fear not, for it can be broken down into digestible steps, making it approachable even for those just dipping their toes into this fascinating field. It's crucial to remember, though, that the method we're about to delve into harnesses man-made energy, which might not be the ideal approach for beginners or those keen on tapping into the bounty of natural energies.

The allure of this process is rooted in its exploratory nature. Each step is a revelation, each observation a window into the intricate interplay between electrical fields and the pulsating life of plants. But let's not forget, safety is of utmost importance. When dealing with the potent force of electricity, always adhere to safety protocols. If any step seems murky or uncertain, don't hesitate to seek guidance from a professional or turn to educational resources such as YouTube tutorials for a more comprehensive understanding.

This method not only offers a unique avenue to boost plant growth and productivity, but it also unravels a deeper understanding of the complex relationship between electricity and life. However, let's not forget that our journey doesn't end here. In the next chapter, we'll venture further into the techniques and tools that harness natural energy, aligning with the principles of permaculture. This is just the beginning of our exploration, a glimpse into the vast expanse of possibilities that electroculture holds.

Collect System Components: The first step is gathering the necessary components. This stage is akin to assembling the pieces of a puzzle. Each component plays a crucial role in the overall system. The power source, which could be a wall charger or battery, provides the electrical energy that will stimulate the plants. The hook-up wire serves as the conduit for this energy, while the electrodes, typically made from iron or steel nails, interface directly with the soil, transmitting the electrical energy into the growing medium. Additionally, you'll need a growing container filled with soil and seeds, ready to receive this energy. To assemble these components, you'll need basic tools like wire strippers and pliers. For those aiming for a more robust setup, a soldering iron can be a valuable addition to your toolkit.

Prepare Power Supply: With all the components at hand, the next step is to prepare the power supply. This stage involves some basic electrical work. The power connector at the end of the cord needs to be removed, and a portion of the plastic insulation stripped away to expose the bare wire. This exposed wire will serve as the connection point between the power supply and the electrodes, bridging the gap between the source of electricity and the soil.

Attach Wires to Electrodes: Once the power supply is prepared, the next step is to attach the wires to the electrodes. This is where the electrical circuit starts to take shape. The wires can be twisted around the electrodes and secured with electrical tape or heat-shrink tubing. If you have a soldering iron, you can use it to solder the wires to the electrodes, creating a more secure and durable connection.

Insert Electrodes into Growing Medium: With the wires securely attached to the electrodes, the next step is to insert the electrodes into the growing medium. This is a critical step, as the placement of the electrodes can significantly impact the effectiveness of the system. The electrodes need to be positioned in such a way that the electric field they generate covers the entire root zone of the plants. This ensures that all parts of the plant can benefit from the electrical stimulation.

Power On and Observe: The final step in setting up an electroculture system is to power on the system and observe. This is where the real magic happens. As the system comes to life, the electrical field begins to interact with the plants, and the effects of this interaction start to become visible. It's important to monitor the plants closely during this stage, observing how they respond to the electrical stimulation. Adjustments may need to be made to the intensity or duration of the electrical stimulation based on these observations.

Setting up an electroculture system is more than just a technical process. It's a journey of discovery and learning. It requires patience, curiosity, and a willingness to experiment. But for those who embark on this journey, the rewards can be immense. Not only does it offer a unique way to enhance plant growth and productivity, but it also provides a deeper understanding of the intricate interplay between electricity and life.

Maintaining an Electroculture System

Maintaining an electroculture system is an ongoing process that requires attention, care, and a keen eye for observation. It's not just about setting up the system and letting it run; it's about continuously monitoring and adjusting the system to ensure optimal performance and plant health. This process can be broken down into several key areas:

Monitoring Plant Health: The health of your plants is the most direct indicator of the effectiveness of your electroculture system. Regularly observe your plants for signs of growth, vitality, and overall health. Look for changes in leaf color, size, and shape, as well as the development of flowers and fruits. Any changes, whether positive or negative, can provide valuable feedback on the performance of your system.

Adjusting Electrical Parameters: The electrical parameters of your system, such as voltage and frequency, may need to be adjusted over time. This could be in response to changes in plant health, changes in environmental conditions, or simply as part of an ongoing process of experimentation and optimization. Always remember to make adjustments gradually and observe the effects carefully.

Maintaining Electrodes and Wiring: The physical components of your system, such as the electrodes and wiring, also require regular maintenance. This can include cleaning the electrodes to remove any buildup of minerals or other substances, checking the wiring for signs of wear or damage, and replacing any components as necessary.

Safety Checks: Safety should always be a top priority when working with electricity. Regular safety checks should be carried out to ensure that all components are functioning correctly and safely. This can include checking for any signs of overheating, ensuring that all connections are secure, and verifying that the system is properly grounded.

Record Keeping: Keeping detailed records of your electroculture system can be incredibly useful. This can include records of any changes made to the system, observations of plant health and growth, and any other relevant data. These records can provide valuable insights into the performance of your system over time and can help guide future adjustments and improvements.

Maintaining an electroculture system is an ongoing journey of observation, adjustment, and learning. It's about more than just keeping the system running; it's about continuously striving to optimize the system and maximize the benefits for your plants. And while it can be a challenging journey, it's also one that can be incredibly rewarding. With each adjustment and observation, you'll gain a deeper understanding of the intricate interplay between electricity and plant life, and with each success, you'll see the tangible results of your efforts in the form of healthier, more vibrant plants.

Chapter 6 - Getting Hands-On: Tools and Materials for Electroculture

As we turn the page to this new chapter, we find ourselves standing at the precipice of an exciting transition. We're moving from the realm of theory into the tangible world of practice, where the fruits of our labor in the field of electroculture begin to bloom. This chapter is a deep dive into the practical aspects, with a particular focus on systems that harness the power of natural energies. It's here that we'll explore the tools and materials that make these systems possible, and guide beginners on how to cultivate their own electroculture garden.

Electroculture is more than just understanding the science behind it or appreciating its benefits. It's about taking that knowledge, rolling up your sleeves, and applying it in a practical way. It's about transforming your garden into a vibrant oasis, pulsating with life, powered by the invisible yet potent forces of natural electricity. To achieve this, you need the right tools and materials.

In the pages that follow, we'll explore some of the common tools used in natural energy-based electroculture. We'll delve into the Spiral, the Pyramid, the Energy Tower, and the Lakhovsky Ring. I'll also provide step-by-step DIY instructions for creating your own tools, empowering you to start experimenting in your own garden.

We'll also discuss the importance of the direction of energy flow in your system. Is it clockwise or anti-clockwise? How does this affect the growth and health of your plants? We'll explore these questions and more. And of course, we'll cover the safety precautions and best practices you need to follow to ensure that your journey is not only successful but also safe.

So, whether you're a seasoned gardener looking to boost your yields, or a curious beginner eager to experiment with this fascinating field, this chapter has something for you. As we delve into the world of natural energy-based electroculture, we're embarking on a journey that promises to be as enlightening as it is rewarding. So, are you ready to get started? Let's dive in!

Common Tools

The various tools that make this innovative approach to agriculture possible. These tools, each with their unique characteristics and functions, are the conduits through which we harness the power of electricity to stimulate plant growth and vitality.

Here's a list of some of the most common tools used:
- **Spiral**
- **Energy Tower / Irish Basalt Tower**
- **Lakhovsky Rings**
- **Pyramid**
- **Magnetic Antennas and Cylinders**
- **Genesa Crystal**
- **Copper coil antennas**

In the following sections, we will delve deeper into each of these tools, exploring their design, function, and how you can use them in your electroculture system. Remember, understanding these tools is just the first step. The real magic happens when you start using them in your garden, observing their effects, and experimenting with different setups.

Materials and DIY Instructions For Creating Your Own Tools

Are you ready to plunge into the heart of electroculture, to feel the pulse of the earth beneath your fingers? This is where our journey takes a hands-on turn, where we transition from observers to active participants. This is where we empower you, dear reader, to craft your own tools of electroculture.

Isn't there something profoundly satisfying about creating with your own hands? It's not just about the cost-effectiveness, although that's a definite perk. It's about the thrill of customization, the joy of tailoring a system to fit your unique needs and preferences. It's about the deep connection that forms when you understand not just the 'what' but the 'how' of electroculture.

In the following pages, we will embark on a journey of creation. I will be your guide, leading you through the labyrinth of materials and methods. Fear not if you're a novice, or if DIY is an alien concept to you. The instructions that follow are designed to be your compass, simple and clear, requiring no prior knowledge or specialized skills.

We will explore the art of crafting a range of electroculture tools, from the elegant spirals to the intricate Lakhovsky rings. But our journey doesn't end there. I will also illuminate the path to positioning these tools for maximum effect.

So, are you ready to seize the power of electroculture, to mold it with your own hands? Let's take that first step together!

SPIRAL

Have you ever marveled at the intricate patterns found in nature? The mesmerizing arrangement of seeds in a sunflower, the spiraling arms of a distant galaxy, the delicate curl of a fern frond as it unfurls. These patterns, governed by the Fibonacci sequence or the Golden Ratio, are not just aesthetically pleasing. They hold a secret, a key to enhancing the growth and vitality of plants. This secret is embodied in a simple tool used in electroculture: the spiral.

Crafted from copper wire, these spirals are more than just a beautiful representation of nature's patterns. They are conduits, gateways that channel the Earth's natural electric and magnetic fields into the soil and the plants. Can you imagine it? The invisible energy of the Earth, flowing through the spiral, infusing the soil and the plants with vitality and life.

The design of these spirals is not a product of chance. It is inspired by the pioneering research of Pierre Luigi Ighina. The spirals are meticulously crafted to direct atmospheric energy towards the base of the cone. When nestled in the soil near a plant, the spiral comes alive, channeling the Earth's energy into the plant, boosting its growth and health.

But what if you're faced with a tree plagued by disease or an unwelcome fungus? The spiral is versatile. Simply place it the other way around, directing its energy upward from the earth. This could provide the much-needed relief for the tree, helping it combat the issues it's facing.

Creating these spirals requires a length of copper wire and a few basic tools. The wire is then shaped into a spiral, a process that is as fascinating as it is rewarding.

The size of the spiral can vary. It needs to be large enough to effectively channel the electric and magnetic fields, yet not so large that it becomes a cumbersome presence in your garden or farm.

It's important to remember that the journey of electroculture is a dance with nature. The effectiveness of a spiral can depend on the rhythm of this dance, the type of soil, the type of plant, the whispers of the local climate. It's a dance that might require some experimentation, some trial and error, to find the rhythm that works best for your specific situation.

If you're looking to create these spirals on a larger scale, consider investing in a mould. And while these spirals can be crafted from any metal, copper often takes the lead. Its excellent conductivity makes it the perfect partner in this dance with nature.

Materials Needed:
1. Copper wire (about 2-3 meters long)
2. Wire cutters
3. A cylindrical object (like a broom handle or a thick dowe or a cylindrical moldl) to shape the wire

Steps to Create a Spiral

1. Cut the Wire: Using the wire cutters, cut a length of copper wire. The length can vary depending on the size of the spiral you want to create, but 2-3 meters is a good starting point.

2. Shape the Wire: Wrap the copper wire around your cylindrical object (or the mold) to create the spiral shape. The spirals should be evenly spaced and the wire should not overlap.

3. Remove the Spiral: Carefully slide the spiral off the cylindrical object. You should now have a copper wire spiral.

4. Install the Spiral: Choose a location near your plants to install the spiral. The spiral can be placed directly on the ground, or it can be elevated using a stake or a similar object. If you're using multiple spirals, they should be spaced evenly throughout your garden.

Buying a Spiral:

If you prefer to buy a spiral instead of making one, there are many options available online and in stores. When buying a spiral, look for one that is made of copper and is designed for use in gardens or farms.

Alternatives

Here there are two alternative methods you can use:

Method 1

1. Gather your materials: You'll need insulated copper wire and pin-nose pliers (If the wire isn't too thick, you can also just use your hands) .

2. Begin by rolling the wire into a circle. Keep the wire pressed firmly against the table and slowly twist it around with your hand. You won't need the pliers for this step.

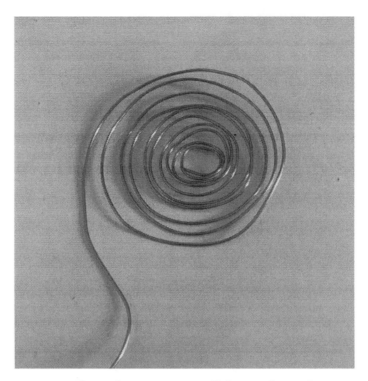

3. Maintain pressure on the wire as you roll it against the outer edge of the circle. The final result should resemble a spiral.

4. Finally, grab the end of the spiral with your pliers and gently pull it outward. This will give your spiral its final shape.

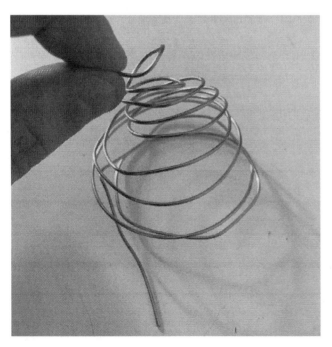
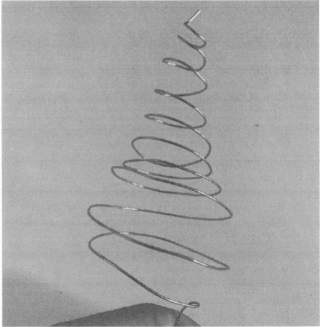

Method 2

You'll be twisting the copper around a funnel. This can help create a more uniform spiral shape.

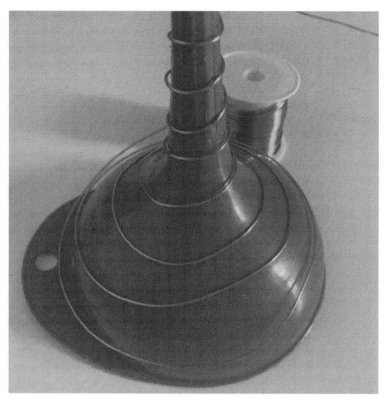

Using a slightly longer funnel than the one shown in this photo would be more ideal.

Personal Safety Precautions:

- **Handling Copper Wire:** Copper wire can have sharp edges, so it's important to handle it carefully to avoid cuts or scratches. Wearing gloves can provide an extra layer of protection.
- **Working with Tools:** When cutting and shaping the copper wire, be sure to use the appropriate tools and follow safety guidelines. Always cut away from your body and keep your fingers clear of the cutting edge.
- **Installing the Spiral:** When installing the spiral in your garden or farm, be mindful of your surroundings. Avoid installing the spiral near power lines or

other electrical equipment to prevent electrical accidents.

Soil Health Precautions:

- **Monitor Soil Health**: Regularly check the health of your soil. Look for changes in soil color, texture, or smell that could indicate a problem. Regular soil testing can also help you keep track of nutrient levels and pH balance.
- **Watch for Signs of Copper Toxicity:** Keep an eye on your plants for signs of copper toxicity, such as stunted growth or discolored leaves. If you notice these signs, consider having your soil tested for excessive copper.
- **Manage Copper Use:** If you're using copper spirals, be mindful of other sources of copper that might be contributing to the total copper load in your soil, such as copper-based fungicides or fertilizers. Too much copper can be harmful to plants and soil organisms.
- **Consider Alternatives**: If you're concerned about copper toxicity, consider using spirals made from other materials. While copper is commonly used due to its high conductivity, other materials like aluminum or galvanized steel can also be used.

ENERGY TOWER / IRISH BASALT TOWER

In the grand tapestry of electroculture, a standout thread is the Energy Tower, also known as the Irish Basalt Tower. This tool is a testament to the power of nature, harnessing the innate properties of basalt, a volcanic rock, to breathe life into plants. Basalt, a rock born from the fiery heart of volcanoes, is a treasure trove of minerals and trace elements. But what makes it truly special is its paramagnetic property. Can you imagine a rock holding an electric charge, cradling it within its stony heart, and then releasing it slowly, like a gentle whisper in the wind? This gradual release of energy, a gift from the basalt to the plants, can work wonders on plant growth.

The Energy Tower, as the name suggests, is a tower-like structure, a monument to the power of nature. It's constructed by stacking basalt stones in a specific configuration, each stone a building block in this tower of energy. Once the tower is built, it's placed in the garden or farm, a silent sentinel amidst the greenery.

Now, let's embark on a journey to create your own Energy Tower. Here's a step-by-step guide to lead you on this path:

Materials Needed:

- Basalt stones: These can be sourced from a local stone supplier or online. The size and number of stones you need will depend on the size of the tower you want to build.

Steps:

- **Select a Location:** Choose a location in your garden or farm for your Energy Tower. The location should be central to the area where you want to enhance plant growth.
- **Prepare the Base:** Clear the area and prepare a stable base for your tower.

This could be a flat stone or a concrete slab.

- **Stack the Stones**: Start stacking the basalt stones on top of each other. The stones should be stacked in a way that creates a stable, tower-like structure.
- **Secure the Tower:** Once the tower is built, it's a good idea to secure it to prevent it from toppling over. This could be done by using a metal rod or wooden stake that runs through the center of the tower.

Buying an Energy Tower:

Purchasing a pre-made Energy Tower can be a convenient option if you're short on time or don't feel comfortable building one yourself. There are several online retailers and garden supply stores that sell pre-made Energy Towers. However, these can be quite expensive due to the cost of the basalt stones and the labor involved in building the tower.

Precautions

- **Stability is Key:** Make sure your tower is stable and secure to prevent it from toppling over and causing injury or damage.
- **Monitor Plant Health:** As with any new addition to your garden or farm, monitor your plants closely to see how they respond to the Energy Tower. If you notice any negative effects, consider moving the tower to a different location or consulting with an expert.

Remember, the goal of electroculture is to work with nature, not against it. By using tools like the Energy Tower, you can harness the Earth's natural energy to enhance plant growth and improve the health and productivity of your garden or farm.

LAKHOVSKY RINGS

In the realm of electroculture, there exists a tool of intriguing potential, a tool that whispers the name of its creator, Georges Lakhovsky. These are the Lakhovsky Rings, also known as Lakhovsky Antennas. Circular bands of copper, designed with a singular purpose - to harness and channel the Earth's natural electromagnetic energy, to breathe life into plants and enhance their growth.

The original design of these rings, as conceived by Lakhovsky, was a symphony of concentric circles, each crafted from a different material. Copper was a key player in this ensemble, but it was not alone. Other metals lent their voices to this chorus, each contributing to the range of frequencies that the ring could interact with.

In the hands of DIY enthusiasts, however, the design of the Lakhovsky Ring has been simplified. Copper, with its excellent conductivity and ease of handling, has become the star of the show. It's as if the other metals have stepped back, allowing copper to take the spotlight.

So, how does one create a Lakhovsky Ring?

Materials Needed:

- **Copper wire**: The thickness of the wire can vary, but a thicker wire will generally be more durable and effective.

Steps:

- **Cut the Wire:** Cut a length of copper wire. The length will depend on the size of the ring you want to create.
- **Form the Ring**: Bend the wire into a circular shape. However, do not close the loop. Instead, leave a small gap between the ends of the wire. VERY SIMPLE!
- **Position the Ring:** Place the ring in your garden or farm with the opening

facing North. The ring should surround the area where you want to enhance plant growth, but it should not touch the plants directly.

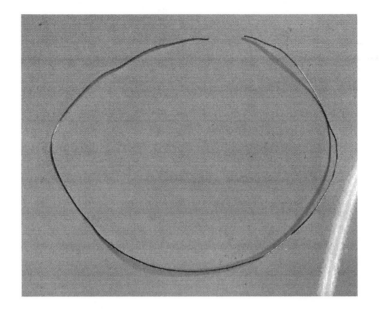

Alternatives

Here's an alternative method for creating a Lakhovsky ring using thin electrical cables:

1. Start by sourcing thin electrical cables. These can often be found inside cables that plug directly into the electricity supply. Once you strip back the black plastic, you should find a minimum of two copper cables.

2. Cut approximately eight cables to your desired size and strip the plastic off the ends.

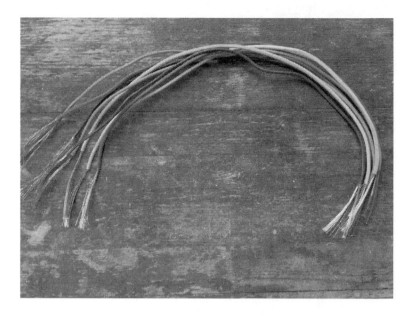

3. Twist the ends of the cables together.

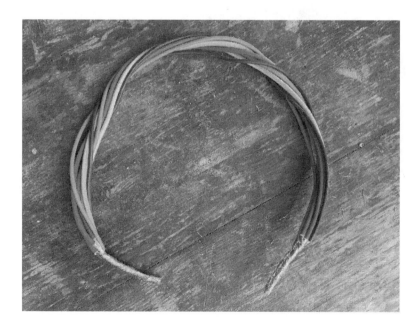

4. You'll need your thin copper wire again at this point to add some stability to your coil.

5. Aim to make the ends (where you stripped the plastic off) dense with copper. This will enhance the effectiveness of your Lakhovsky ring.

Precautions:

- **Avoid Direct Contact:** Copper can potentially be toxic to plants if it comes into direct contact with them. Therefore, it's important to ensure that the

Lakhovsky Ring does not touch your plants directly.
- **Monitor Plant Health**: As with any new addition to your garden or farm, monitor your plants closely to see how they respond to the Lakhovsky Ring. If you notice any negative effects, consider moving the ring to a different location or consulting with an expert.

Lakhovsky Rings are a simple and affordable tool that can be used to enhance plant growth. By harnessing the Earth's natural electromagnetic energy, these rings can help to create a more productive and healthy garden or farm. If you're interested in a more complex build, I would recommend researching Georges Lakhovsky's original designs or consulting with an expert in the field.

PYRAMID

Pyramids have been discovered to positively influence plant growth when utilized in electroculture.

These structures, when constructed with precision, are thought to capture and distribute the Earth's natural electromagnetic energy, enhancing plant health and productivity.

Among all the tools, this one involving pipes requires a bit of dexterity. If you're not comfortable using these tools, I would advise against trying to avoid any injuries. Personally, I prefer using the other tools.

Materials Needed:
- Copper Tubes: Copper is the preferred material due to its superior conductivity and ease of handling. (Wood is also an alternative for some)
- Rivets or copper screws + Riveter / alternatively screws and bolts or copper wires
- Vise, saw, 5/32 inches drill bit

Copper wires can also be used in place of tubes, but always adhere to the measurements.

Procedure:
1. Initially, decide on the type of Pyramid: Cheops or Nubian

2. Now, choose the size which will be dependent on your specific requirements and available space. You will need 8 pieces to construct the Pyramid.

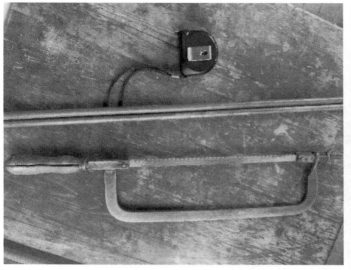 

For this demonstration, I used some fairly old copper tubes, but that's not a problem.

3. However, the ratio between the pyramid's height and the base side must be maintained.

4. In the case of the Cheops pyramid, the slabs used for the height must be 0.952 relative to the base. For the Nubian pyramid, the ratio must be 1.618033

5. Example Cheops : 4 pieces of 10inches and 4 pieces of 9.52 inches (10*0.952)

6. Nubian example: 4 pieces of 10inches and 4 pieces of 16.18033inches

7. Use the hacksaw to cut the 8 pieces according to the measurements obtained.

8. Mark a distance of 1inch from the edge on each piece.

9. Clamp the tube to flatten it, relative to the marked 1inch.

10. Mark a point 1/4 inches from the edge, which will be used to make the 5/32 inch connecting holes.

11. Drill all 8 pieces.

As you can see, I made some holes using a simple drill (only use it if you're confident in your ability to handle it safely to avoid injuries).

12. The sides used for the pyramid's height must be placed in a vise and worked by folding them, to create a corner which facilitates assembly.

13. Assemble with screws or copper wires.

This procedure is just one of the ways in which the pyramid can be assembled. You are free to be creative.

Pyramid Placement: Position the pyramid in your garden or farm. The pyramid should encompass the area where you want to promote plant growth, but it should not directly touch the plants.

Alternatives

A simple alternative would be to use copper wires and find a way to bind them together, always adhering to the specified measurements. (Before constructing your pyramid, take measurements around the plant where you plan to place it.)

Guidelines for Optimal Yield:
1- Orient the 4 sides of the base according to the cardinal points.
2- Insert a spiral at the vertex to increase energy input.
1. Precautions:
3- Avoid sowing during these lunar phases: Apogee / ascending or descending node – Perigee.

Precautions:
- **Avoid Direct Contact**: As with the Lakhovsky Rings, copper can potentially be toxic to plants if it comes into direct contact with them. Therefore, it's important to ensure that the pyramid does not touch your plants directly.
- **Monitor Plant Health:** Monitor your plants closely to see how they respond to the pyramid. If you notice any negative effects, consider moving the pyramid to a different location or consulting with an expert.
- **BE SAFE:** As I've mentioned before, this variant involves the use of slightly more advanced tools, which could pose a challenge for those with less manual dexterity. If this is your first time using these tools, please exercise extreme caution. The instructions provided here are solely for informational purposes, to demonstrate how I constructed my pyramids.

Pyramids can be used for various purposes in electroculture. They can help energize seeds, resulting in healthier and better-tasting plants. They can also be used to fertilize large portions of soil and even aid in meditation. However, to benefit from such a system, it's important to know how to use such a pyramid effectively.

While it's possible to construct your own pyramid, it's also possible to purchase pre-made pyramids designed for use in electroculture. These can be a good option if you're not comfortable with DIY projects or if you want to ensure that your pyramid is built to the correct specifications.

MAGNETIC ANTENNAS AND CYLINDERS

The tool in question is a Magnetic Antenna for ElectroCulture, an invention of Yannick Van Doorne.

This tool is designed to enhance plant growth by harnessing the Earth's natural magnetic fields.

Materials Needed:
- Magnets with a hole: Six magnets are typically used, but the size and number can be adjusted based on availability and specific needs. Magnets can be sourced from various places, including old loudspeakers or refrigerator magnets.
- Wire: The wire is used to connect the magnets and transmit the magnetic field. Copper is not recommended for this tool; instead, galvanized steel wire is preferred due to its ferromagnetic properties.
- Beeswax: This is used to coat the magnets, enhancing their effects by transmitting the beneficial frequencies of the beeswax molecules to the wire.

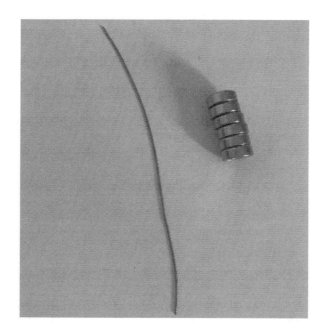

Steps to Build:

1. Gather six magnets and arrange them in a cylinder with a hole in the middle. This arrangement makes it easier to connect the wire.
2. Ensure the magnets are in the correct orientation to maximize their effects.
3. Coat the magnets in beeswax. This step is optional but recommended as it can enhance the effects of the magnets.
4. Connect the wire to the magnets, ensuring good contact.
5. Install the magnetic antenna in your garden or farm

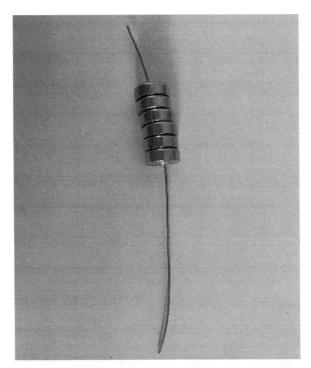

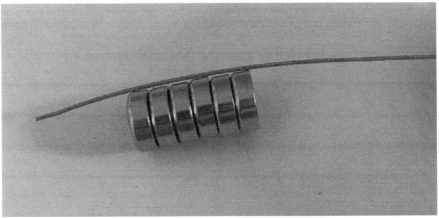

Alternative if the magnets don't have a hole

General Guidelines:

- The magnetic antenna should be installed in a north-south direction, in line with the Earth's magnetic field. This can be determined using a compass.
- The antenna should be installed in a way that it remains stable and does not move around. This can be achieved by making the antenna longer or by using a piece of iron to ensure it stays in place.
- The antenna can be installed either above or below the soil. If installing below the soil, ensure that it is not too deep (maximum 50 cm) as this could affect its effectiveness.
- The magnetic antenna can be used in conjunction with other farming techniques, including organic farming and permaculture, to enhance their effectiveness.
- If you don't have cylinder magnets with holes in them, you can also stick wire to the outside of the cylinders

Precautions:

- If the soil is being plowed, care should be taken to ensure that the antenna is not disturbed. If deep plowing is being done, this technique may not be suitable.
- While iron is used in the construction of the antenna, it should be noted that excessive iron can disturb the Earth's magnetic field. Therefore, it is important to ensure that the antenna is installed correctly and remains stable.
- Over time, the beeswax may decompose or melt, especially in hot climates.

However, this should not significantly affect the functioning of the antenna. Even without the beeswax, the magnets will still have a beneficial effect.

- It's important to note that the effectiveness of the magnetic antenna can vary and may be influenced by various factors, including the specific conditions in your garden or farm and the types of plants being grown.

GENESA CRYSTAL

A Genesa Crystal is a three-dimensional sculpture typically made from copper wire. It is designed to represent the cellular pattern of an embryonic living organism. The Genesa Crystal is a spherical cube octahedron, a shape that is believed to have powerful energetic properties.

Materials Needed:

Copper Wire: The Genesa Crystal is typically made from copper wire, which is known for its excellent conductivity properties.

Steps:

Creating a Genesa Crystal can be quite complex due to its intricate design. It involves bending and shaping the copper wire to create the three-dimensional shape of the Genesa Crystal. This process can be quite challenging, especially for beginners, and may require a significant amount of time, patience, and precision.

Given the complexity of creating a Genesa Crystal, it might be more practical to purchase one. There are various online platforms and stores where you can buy a professionally made Genesa Crystal. This way, you can ensure that the crystal is perfectly constructed to harness its full potential.

Placement of the Genesa Crystal:

- The Genesa Crystal should be placed in a location where it can interact optimally with the electromagnetic fields in the environment. This is because the Genesa Crystal works by interacting with these fields to enhance the growth and health of plants
- Location: The crystal should ideally be placed in the center of your garden or

the area where you're practicing electroculture. This central location allows the crystal to interact with the electromagnetic fields evenly across the entire area.

- Height: The height at which you place the Genesa Crystal can also impact its effectiveness. It should be placed at a height that allows it to interact with the electromagnetic fields without being obstructed by other objects.
- Orientation: The orientation of the Genesa Crystal can also play a role in its effectiveness. Some practitioners suggest aligning the crystal with the magnetic north for optimal results, although this is a topic of ongoing debate in the electroculture community
- Surrounding Environment: The Genesa Crystal should be placed in an environment that is free from electronic interference. Electronic devices can disrupt the electromagnetic fields, reducing the effectiveness of the crystal.
- Remember, the Genesa Crystal is a delicate structure and should be handled with care during placement. It's also a good idea to periodically check the crystal to ensure it remains in the optimal position and condition.

In conclusion, the Genesa Crystal is a unique and powerful tool in electroculture. While it may be more complex to create than some of the other tools we've discussed, the potential benefits make it well worth the effort. If you're a beginner, purchasing a pre-made Genesa Crystal could be a more feasible option.

COPPER COIL ANTENNAS

In the context of electroculture, Copper coil antennas take on an additional role as conductors of electricity, helping to distribute the beneficial effects of electric fields to the plants.

Materials Needed:
- Sticks: These should be naturally fallen sticks from your backyard. The idea is that these sticks are at the same frequency as you and your plants, which can enhance the effectiveness of the antenna.
- Copper Wire: Copper wire is used because of its excellent electrical conductivity properties.
- Wire Cutters: These are used to cut the copper wire to the desired length.

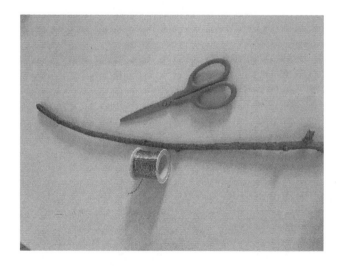

Steps to Build:
1- Collect some naturally fallen sticks from your backyard.
2- Cut a length of copper wire using the wire cutters.
3- Spiral the copper wire up the sticks to create your antennas. The spiraling of the wire around the stick is what creates the antenna effect. You can create a spiral that extends from the stick by adhering to the instructions provided

above (refer to the instructions in Alternative Method 1)

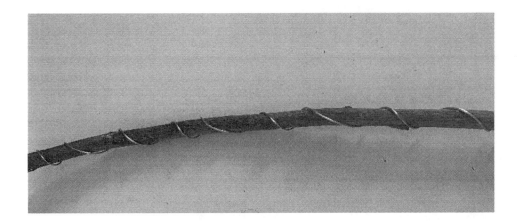

4- Place your antenna into your plant's pot, raised bed, garden, farm, or even in your basement. The antenna can be used with any type of plant, including potatoes.

General Guidelines:
- The atmospheric antenna works by increasing the electrical conductivity of the soil, which can help your plants grow faster.
- The antenna can also make your plants more resistant to heat and frost and

reduce their water needs.
- The antenna can be used in conjunction with other farming techniques, including organic farming and permaculture, to enhance their effectiveness.

Precautions:
- Be careful when handling the copper wire and using the wire cutters to avoid injury.
- Ensure that the antenna is securely placed in the soil and will not pose a tripping hazard or other safety risk.
- Monitor your plants regularly to assess the effectiveness of the antenna and make any necessary adjustments.

Variation

Alternatively, you could extend the spiral beyond the end of the stick. You could also wrap additional copper wire around the stick and continue the copper wire beyond the end of the stick, creating actual antennas. There isn't a single correct method for constructing these antennas. The best approach is to experiment and test different configurations.

Choosing the Right Tool for you

Now that we've explored the various tools used ,it's time to consider which one might be the best fit for your specific needs. Each tool has its unique features and benefits, and understanding these can help you make an informed decision. Whether you're a beginner just starting out or an experienced practitioner looking to experiment with new techniques, the right tool can make a significant difference in your electroculture journey. In this section, I'll guide you through the process of choosing the right tool for your garden or farm, taking into account factors such as the type of plants you're growing, the size of your garden, and your specific goals for using electroculture. Let's dive in!

1. **Copper coil antennas**: These are the simplest tools to start with. They are easy to install and can be used with any type of plant. If you're a beginner and just starting to experiment with electroculture, copper coil antennas are a great starting point. They can be used in any size of garden, from a small home garden to a larger farm.

2. **Lakhovsky Rings:** These are also quite simple to use and are a good choice for beginners. They can be used with any type of plant and are easy to install. They're particularly useful if you're interested in enhancing the growth of individual plants or small groups of plants.

3. **Spiral:** The spiral is a bit more complex than the copper coil antennas or Lakhovsky rings, but it's still quite accessible for beginners. It's versatile and can be used in a variety of settings. If you're ready to experiment a bit more and want to enhance the growth of individual plants or small groups of plants,

the spiral could be a good choice.

4. **Magnetic Antennas and Cylinders:** These tools are a little more advanced and require a more detailed understanding of electroculture principles. They can be a good choice for those ready to take their experiments to the next level.

5. **Pyramid:** The pyramid is a more specialized tool. It requires careful placement and orientation to be effective, but it can be a good choice if you're growing specific types of plants that respond well to the type of field created by the pyramid.

6. **Energy Tower/Irish Basalt Tower:** This tool is more advanced and typically better suited for those who have already gained some experience with electroculture. While it requires a more detailed understanding of electroculture principles, it's often more practical to purchase this tool rather than attempting to construct it yourself. Once acquired, its placement is not particularly challenging, making it a feasible option for those ready to delve deeper into the world of electroculture.

7. **Genesa Crystal:** This tool is quite challenging to construc (you can prefer to buy it). However, once created, its placement is relatively straightforward. It's an excellent choice for those who have a solid grasp of electroculture and are eager to explore its full potential.

Remember, the best tool for you depends on your specific needs and circumstances. Start with the simpler tools and then experiment with more advanced tools as you gain experience and confidence with electroculture. Always ensure safety when installing and using these tools, and don't hesitate to seek advice from more experienced practitioners or professionals if needed.

Safety Precautions and Best Practices

It's essential to take a moment, a brief interlude, to reflect on the importance of safety precautions and best practices. We've ventured into the realm of various tools and systems, each with its unique allure and benefits. Yet, no matter which path you choose to tread, certain principles stand as unwavering beacons.

Consider the materials that form the backbone of your electroculture system. Picture a Lakhovsky ring or spiral, gleaming in the sunlight, its copper surface a testament to its role in this grand symphony of growth. Yet, like the most enchanting of melodies, it carries a note of caution. Copper, while a vital nutrient for plants in moderation, can transform into a silent saboteur in excess. Can you see the telltale signs of copper toxicity, the leaves of your plants discoloring like a sunset sky? If copper tools are your companions in this journey, vigilance must be your watchword.

As we navigate the intricacies of electroculture, we also encounter practices that can amplify the effectiveness of our efforts. Imagine the soil in your garden, a rich tapestry of life and nutrients. Can you feel its moisture, a vital conduit enhancing the electric field's reach? Dry soil, in contrast, is like a broken bridge, hindering the flow of conductivity. Can you see the importance of maintaining this delicate balance?

By adhering to these safety precautions and best practices, you can ensure a safe and effective electroculture experience. As we move forward, we'll continue to explore the practical applications of these systems and tools, and how to best utilize them in your gardening endeavors.

Clockwise or anti-clockwise?

The question of whether to coil electroculture devices clockwise or anti-clockwise is a common one among beginners in the field. The direction of the coil can influence the energy flow and, consequently, the effect on the plants. But it's important to clarify that the direction of the coil is determined when looking at the device from above, not from below.

In the Northern Hemisphere, it's generally recommended to coil the devices in a clockwise direction when viewed from above. This is because the Earth's magnetic field in the Northern Hemisphere directs the magnetic field lines from the North to the South. Coiling the devices clockwise aligns with this natural direction, potentially enhancing the effectiveness of the device.

Conversely, in the Southern Hemisphere, the Earth's magnetic field lines direct from South to North. Therefore, when viewed from above, it's suggested to coil the devices in an anti-clockwise direction to align with the natural direction of the magnetic field.

However, these are general recommendations and the effectiveness can vary depending on other factors such as the type of plant, the specific electroculture system being used, and the local environmental conditions.

Some practitioners also believe that the direction of the coil might influence the type of energy being harnessed – clockwise for positive or 'masculine' energy, and anti-clockwise for negative or 'feminine' energy. This is more aligned with esoteric and spiritual beliefs and less with scientific understanding, but it's a perspective that some electroculture practitioners adhere to.

In conclusion, while the direction of the coil can potentially influence the effectiveness of the electroculture device, it's just one of many factors to consider. It's always recommended to experiment and observe the effects on your specific plants and in your specific environment. After all, the beauty of electroculture lies in its potential for customization and adaptation to individual needs and conditions.

Chapter 7 - Harnessing the Power of Electroculture: Practical Applications

This chapter is where theory meets practice. As we embark on the seventh chapter of our exploration, we find ourselves standing at the precipice of practicality. We've traversed the historical landscapes of electroculture, delved into the scientific underpinnings, and navigated the tools and systems that make it possible. Now, it's time to witness the alchemy of these elements in the crucible of real-world applications.

This chapter is the bridge between the abstract and the concrete. We will probe the myriad ways electroculture can be wielded, from the stewardship of soil health to the enhancement of seed germination and crop yield.

Our first stop is soil management, the bedrock of any agricultural endeavor, and even more so in electroculture. We'll unearth the ways electroculture can bolster soil health and fertility, and the optimal practices for managing soil within an electroculture framework.

From the soil, we turn our attention to the miracle of seed germination. We'll delve into the ways electroculture can amplify this process, fostering the growth of healthier, more robust seedlings. Practical tips on leveraging electroculture for seed germination will also be shared, providing a roadmap for those eager to apply these insights.

The journey continues with a deep dive into crop growth and yield, another significant boon of electroculture. We'll discuss how electroculture can invigorate plants, accelerating their growth and enhancing their health, leading to bountiful yields. Practical tips on harnessing electroculture to boost crop growth and yield will also be shared, offering a practical guide for those ready to put these insights into action.

Soil Management

In the realm of agriculture, soil management is a cornerstone, and electroculture is no exception. As we've journeyed through previous chapters, we've come to understand that soil is the bedrock of agriculture. It is the nurturing cradle where plants draw their nutrients, and the conduit through which they absorb life-giving water. In the world of electroculture, soil takes on an additional role, serving as the conductor for the electrical currents that invigorate plant growth.

When we delve into the specifics of soil management within the context of electroculture, several key considerations come to the fore. The first of these revolves around the soil's ability to conduct electricity effectively. This capability is influenced by factors such as the soil's moisture content, its mineral composition, and its pH level. Consider, for instance, a patch of dry soil. Its ability to conduct electricity is limited, underscoring the importance of maintaining optimal soil moisture. Similarly, soils rich in minerals are more efficient conductors of electricity compared to their mineral-deficient counterparts.

Moving on to the second consideration, we find ourselves in the microscopic world of soil microorganisms. These minute creatures are the unsung heroes of soil health, tirelessly breaking down organic matter and making nutrients accessible to plants. Intriguingly, some research suggests that electrical currents may stimulate the activity of these microorganisms, potentially enhancing soil fertility. However, this is a field where the final word has not yet been spoken, and more research is needed to fully comprehend these effects and how to harness them optimally.

The third aspect of soil management in electroculture concerns the potential effects of electrical currents on soil pH. Some studies hint at the possibility that electrical currents could alter soil pH, making it more acidic or alkaline. This is a significant consideration as different plants have different pH preferences, and any shift in soil pH could impact plant health and growth. Consequently, regular soil testing and pH adjustments may become a necessary part of managing an electroculture system.

In essence, soil management in electroculture is a delicate dance that combines traditional soil management practices with considerations unique to the use of electrical currents in agriculture. By gaining a deep understanding of these factors and managing them effectively, farmers and gardeners can unlock the full potential of electroculture, paving the way for healthy, productive plant growth.

Seed Germination

The journey of a seed, from its dormant state to a thriving plant, is a marvel of nature. It's a critical phase in the plant's life cycle, a phase where electroculture can play a pivotal role. Picture a seed, nestled in the nurturing embrace of the earth, beginning to absorb water as it finds itself in a conducive environment. This absorption triggers a cascade of events within the seed. Enzymes spring into action, breaking down the stored food reserves and setting the stage for the emergence of a new plant.

Now, imagine the role of electroculture in this intricate dance of life. The electric fields, generated by electroculture systems, can act as a catalyst, accelerating the absorption of water by the seed. How does this happen, you ask? Water, being a polar molecule with a positive and a negative end, can be aligned by the electric field in a way that facilitates its penetration through the seed coat. This hydration of the seed is the first step towards its transformation into a plant.

But the magic doesn't stop there. It also enhances the activity of the enzymes that orchestrate the breakdown of the seed's food reserves. These enzymes, which are proteins and hence polar molecules, can be aligned by the electric field to perform their task more efficiently. This results in a quicker and more effective breakdown of food reserves, fueling the growth of the new plant.

Electroculture also plays a crucial role in guiding the growth of the root and shoot systems. The electric field can steer these systems, ensuring they grow in the right direction and at the right pace. The result? A robust, healthy plant that is well-equipped to thrive in its environment.

However, it's important to remember that electroculture is not a one-size-fits-all solution. The response of seeds to electroculture can vary, influenced by factors such as the type of seed, the strength of the electric field, and the specific environmental conditions. Therefore, it's crucial to experiment with different settings and conditions to find the sweet spot for your specific situation.

So, how can you harness the power of electroculture for seed germination? It can be as simple as placing your seeds near an electroculture device before planting them. You could, for instance, place your seeds near a Lakhovsky ring or a spiral for a few hours or days before planting. Alternatively, you could establish an electroculture system in your garden or greenhouse and plant your seeds directly under it. The world of electroculture is yours to explore, and the journey begins with a single seed.

Improving Crop Growth and Yield

In the verdant world of electroculture, the ultimate quest is to enhance the growth and yield of crops. But how do we achieve this? The answer lies in the power of electricity, a force that can stimulate plant growth and amplify their natural processes. The exploration of electroculture as a tool to boost crop growth and yield is an exhilarating frontier, one that is steadily gaining momentum in the agricultural sphere.

At the heart of electroculture lies the concept of electric fields, a silent yet powerful force that influences plant growth. Imagine a field of corn, each stalk reaching towards the sky, not just driven by sunlight, but also by the invisible pull of an electric field. This field, gentle in its intensity, doesn't harm the plants. Instead, it whispers to them, a subtle stimulus that enhances their growth and yield.

In electroculture, the electric fields are typically applied to the soil or growing medium. This can be done using various methods, such as using electrodes inserted into the soil or using specially designed electroculture devices. The electric fields can stimulate the activity of beneficial soil microorganisms, enhance nutrient uptake by the plants, and stimulate plant growth processes.

One of the key benefits of electroculture is its potential to increase crop yield. By enhancing plant growth and nutrient uptake, electroculture can help plants to produce more fruits, vegetables, or grains. This can be particularly beneficial for farmers, as it can help to increase their productivity and profitability.

Electroculture also weaves a protective shield around the crops, bolstering their resilience. It arms them against diseases and pests and fortifies them against environmental stresses like drought or frost. This ensures a more consistent and reliable crop yield.

In conclusion, electroculture is a beacon of promise in the quest to improve crop growth and yield. It harnesses the power of electricity to enhance plant growth, increase crop yield, and improve the quality of crops. It's a testament to the power of innovation and the endless possibilities that lie at the intersection of agriculture and electricity.

Chapter 8 – FAQs About Electroculture

Embarking on the captivating voyage through the realm of electroculture, it's only natural for questions to arise. This chapter is a beacon, illuminating the path for those who find themselves with queries or uncertainties. Whether you're a novice just dipping your toes into the waters of electroculture, or a seasoned gardener seeking to broaden your horizons, this chapter is designed to dispel common misconceptions and provide clarity.

I've strived to cover a broad spectrum throughout this book, yet this chapter serves as a lighthouse, guiding you back to the answers you seek. It's a quick reference guide, a compass you can turn to whenever you find yourself adrift in a sea of questions.

So, let's delve into the inquiries that many enthusiasts, just like you, often ponder. Let's unravel the mysteries of electroculture together, one question at a time.

1. **What is electroculture?**
   Electroculture is a technique that uses electric and magnetic fields to stimulate plant growth and increase crop yields. It's a natural and environmentally friendly method that can be used in combination with traditional gardening practices. For a more detailed explanation, refer to Chapter 1.

2. **How does electroculture work?**
   Electroculture works by harnessing the natural electric and magnetic fields present in the environment. These fields are manipulated using various tools and devices to create conditions that promote plant growth. You can find a comprehensive explanation in Chapter 2.

3. **What are the benefits of electroculture?**

   Electroculture offers numerous benefits, including faster plant growth, healthier plants, increased crop yields, and enhanced flavor of fruits and vegetables. We delve into these benefits in Chapter 4.

4. **Can electroculture harm my plants?**

   When used correctly, electroculture should not harm your plants. However, it's important to follow the recommended safety precautions and best practices, which we discuss in Chapter 6.

5. **What types of plants can benefit from electroculture?**

   A wide variety of plants can benefit from electroculture, including vegetables, fruits, and ornamental plants. We cover this topic extensively in Chapter 3.

6. **How do I set up an electroculture system in my garden?**

   Setting up an electroculture system involves installing specific tools and devices in your garden. These can range from simple tools like Lakhovsky rings to more complex systems like energy towers. For a step-by-step guide on setting up an electroculture system, refer to Chapter 5.

7. **What materials do I need to start with electroculture?**

   The materials you need will depend on the type of electroculture system you plan to set up. We provide a detailed list of materials and instructions for creating your own tools in Chapter 6.

8. Is electroculture suitable for beginners?

Absolutely! Electroculture can be as simple or as complex as you want it to be. We've designed this book to be accessible for beginners, with clear explanations and step-by-step guides. Check out Chapter 6 for beginner-friendly electroculture systems.

9. Can electroculture be combined with other gardening methods?

Yes, electroculture can be used in conjunction with other gardening methods, including organic farming and permaculture. It's a versatile technique that can enhance the effectiveness of other practices. For more information on this, refer to Chapter 3.

10. What is the science behind electroculture?

Electroculture is based on the principle that electric and magnetic fields can influence plant growth. This involves concepts from physics, biology, and earth science. We delve into the science behind electroculture in Chapter 2.

11. Does electroculture require a lot of maintenance?

The level of maintenance required for an electroculture system depends on the complexity of the system. Some systems, like Lakhovsky rings or spirals, require minimal maintenance, while others may require more regular checks.

12. Can electroculture help with pest control?

There is evidence to suggest that electroculture can help make plants more resistant to pests and diseases. However, it's not a substitute for good gardening practices and should be used as part of an integrated pest management strategy. For more on this, see Chapter 4.

13.   How long does it take to see results with electroculture?

The time it takes to see results with electroculture can vary depending on a number of factors, including the type of plants you're growing, the specific electroculture system you're using, and the growing conditions. However, many gardeners report seeing improvements within a few weeks to a few months. We discussed this in Chapter 7.

14.   Can I use electroculture in a greenhouse?

Yes, electroculture can be used in a greenhouse. In fact, a controlled environment like a greenhouse can make it easier to manage the electric and magnetic fields. For more on this, refer to Chapter 5.

15.   Do I need any special training to use electroculture?

No special training is required to use electroculture. This book provides all the information you need to get started, from understanding the basic principles to setting up your own electroculture system. If you're new to electroculture, we recommend starting with Chapter 1 and working your way through the book.

16.   What are the costs associated with electroculture?

The costs can vary greatly depending on the complexity of the system you choose to implement. Some simple systems, like Lakhovsky rings or spirals, can be made at home with inexpensive materials. More complex systems may require a larger investment. We discussed this in more detail in Chapter 6.

17.   Can I use electroculture to grow specific types of plants?

Electroculture can be used to grow a wide variety of plants, from fruits and vegetables to flowers and trees. However, different plants may respond differently to electroculture, and some may benefit more than others.

OTHER FAQs

1. Is it possible to substitute a copper pipe for an antenna?

While a copper pipe can be utilized, copper coils tend to yield superior outcomes due to their ability to channel energy flow.

2. Can I simply wrap my plants in copper instead of creating an electroculture antenna?

Wrapping plants in copper can be problematic as not all plants tolerate being encased. It's more advisable to construct a basic antenna and position it close to the plants you wish to assist.

3. What should be the height of the electroculture antenna?

The height of atmospheric antennas can be adjusted to your preference. However, antennas that are 6 feet or taller are generally more effective in capturing atmospheric energy.

4. What should be the orientation of my electroculture antenna?

In the Northern hemisphere, your antenna should be wound clockwise, while in the Southern hemisphere, it should be wound counter-clockwise. The winding of the coils should start from the bottom and proceed to the top.

5. **Should I worry about lightning strikes on my antennas?**
   Historically, many old-world structures had numerous lightning rods to stabilize the atmosphere. Therefore, there's no cause for concern.

6. **What is the coverage area of an electroculture antenna?**
   Typically, a 6-foot antenna can cover approximately 225 square feet.

7. **Is electroculture effective on indoor or potted plants?**
   Indeed, electroculture is highly beneficial for indoor plants. A simple antenna can be created using a chopstick and copper for indoor use.

8. **Should the copper of my electroculture antenna be in contact with the soil?**
   Yes, one end of the copper should be in the soil to draw energy from the earth, while the other end collects energy from the atmosphere.

9. **Can I use coated copper or scrap copper wire for electroculture?**
   Yes, you can. If the wire is plastic-coated, ensure to strip the ends so that the bare copper can be inserted into the soil.

10. **Can I use electroculture in my raised beds?**
    Absolutely, just install one antenna in each raised bed. Given that the beds are elevated, the antenna can be between 3 to 6 feet tall.

11.   Is aluminum suitable for use in electroculture?

It's not advisable to use aluminum in the soil due to various health-related concerns.

12.   Does the gauge of the copper wire affect electroculture?

Any gauge of copper wire is acceptable, but a heavier gauge can be used if preferred.

13.   Can I incorporate quartz or crystals into my electroculture antenna?

Definitely! Crystals and gems can emit different color spectrums to stimulate plant growth and attract local pollinators.

14.   Can I use pennies in my garden for electroculture?

Pennies can significantly enhance plant growth, but ensure they were minted before 1980. Most pennies minted in 2023 are composed of 99% zinc and 1% copper.

15.   Can I use electroculture in hydroponics?

Electroculture can be beneficial in hydroponics, but copper pyramids should be used instead of electroculture antennas.

16.   Can I use electroculture for sprouting seeds?

Absolutely! Electroculture can enhance seed sprouting. A simple toothpick wrapped in copper can be highly effective."

As we wrap up this chapter, it's important to remember that the field of electroculture is vast and ever-evolving. The questions addressed here represent some of the most common inquiries, but they are by no means exhaustive. As you continue your journey in electroculture, you may encounter new questions and challenges. That's part of the learning process, and it's what makes this field so exciting and rewarding.

Remember, the answers provided in this chapter are meant to serve as a starting point. They offer a basic understanding and clarification of common doubts. Electroculture is a practice of patience, observation, and continuous learning. As you gain more experience, you'll find your own answers and perhaps even develop new questions. That's the beauty of this journey. So, keep exploring, keep experimenting, and most importantly, keep growing.

Chapter 9 - Gazing into the Crystal Ball: The Future of Electroculture

As we embark on the ninth chapter of our exploration into the captivating world of electroculture, we find ourselves on the brink of tomorrow, peering into the vast expanse of the future. This chapter, aptly titled "Gazing into the Crystal Ball: The Future of Electroculture," is a voyage into the unknown, a journey into the possibilities that lie ahead.

In the chapters that have unfolded so far, we have journeyed through the scientific underpinnings of electroculture, the arsenal of tools and techniques it employs, and the practical applications that bring this intriguing field to life. Now, the time has come to cast our gaze forward, to speculate, to dream about what the future might hold. Like any vibrant field of study, electroculture is a living, breathing entity, constantly evolving and growing. The relentless march of research, the leaps in technological advancements, and the birth of innovative practices are ceaselessly pushing the envelope, redefining the boundaries of what we believe is possible.

Whether you're a seasoned practitioner, with the soil of experience under your nails, or a curious beginner, with the seed of interest taking root in your heart, there's always something new to learn, something new to discover in this ever-evolving field. So, let's turn the page, let's step into the future, and let's see what it holds for us.

Emerging Trends and Innovations

Gazing into the future, the landscape of electroculture stretches out before us, gleaming with promise and brimming with potential. It's a field where innovation sprouts like seeds in fertile soil, each new development pushing the boundaries of what we thought possible. In this little chapter, we'll traverse this exciting terrain, exploring the emerging trends and innovations that are shaping the future of electroculture, while also acknowledging the challenges and opportunities that lie in wait.

One of the most thrilling developments on the horizon is the marriage of electroculture and technology. Picture this: advanced sensors, automated systems, data analytics, and machine learning, all converging to revolutionize electroculture. These technological advancements are not only enhancing the efficiency and effectiveness of electroculture but are also broadening its reach, making it accessible to a wider audience.

Imagine sensors and automated systems working in harmony to provide precise control over the electrical stimulation administered to plants. This precision optimizes plant growth and yield, mitigates the risk of overstimulation, and reduces the need for manual intervention. Now, add data analytics and machine learning to the mix. These tools can help us unravel the intricate dance between electricity and plant growth, leading to more effective and efficient electroculture practices.

Yet, the integration of technology is not without its challenges. It demands a certain level of technical knowledge and expertise, which could be a stumbling block for some. The cost of advanced technology could also be a deterrent, particularly for small-scale farmers and gardeners. And then, there's the risk of becoming overly reliant on technology, which could create a chasm between us and the natural processes that are the lifeblood of agriculture.

As you have already read in this book, another trend that's gaining momentum in the world of electroculture is the shift towards sustainable and organic farming practices. As the environmental repercussions of conventional agriculture become increasingly apparent, many farmers and gardeners are turning to electroculture as a greener alternative. By enhancing plant growth and yield without resorting to harmful chemicals, electroculture could pave the way for a more sustainable and resilient food system.

However, integrating electroculture into organic farming is not without its hurdles. There's a pressing need for more research and concrete evidence to validate the effectiveness of electroculture in organic farming.

Despite these challenges, the future of electroculture is teeming with potential. With ongoing research and innovation, electroculture could play a pivotal role in tackling some of the most urgent issues of our time, from food security and climate change to biodiversity loss and soil degradation.

Potential Challenges and Opportunities

As we cast our gaze towards the horizon, the future of electroculture unfurls before us, a landscape marked by both challenges and opportunities. As we venture deeper into this uncharted territory, we will undoubtedly encounter hurdles that will test our resolve. Yet, it is within these challenges that the seeds of innovation and growth are sown.

One of the most formidable challenges we face is the veil of obscurity that shrouds electroculture. Despite its potential to revolutionize farming, electroculture remains an enigma to many farmers and gardeners. This lack of awareness is a stumbling block to the widespread adoption of electroculture practices. To surmount this challenge, we must invest in education. We need comprehensive resources and training programs that can demystify electroculture and empower farmers and gardeners with the knowledge to harness its potential.

Yet, the path of discovery is not without its twists and turns. Our understanding of electroculture, while growing, is far from complete. There are still vast tracts of unexplored territory, questions that remain unanswered, mysteries that are yet to be unraveled. This underscores the need for continued research and development, for it is through the crucible of scientific inquiry that we can illuminate the unknown.

But let us not lose sight of the opportunities that lie within these challenges. Electroculture, with its promise of healthier plants and higher yields, could be the panacea for many of the ailments plaguing modern agriculture. Imagine a world where farmers are no longer at the mercy of unpredictable weather or ravaging diseases, a world where expensive fertilizers are a thing of the past. Electroculture could make this world a reality, helping farmers increase their profits while reducing their dependence on chemicals.

Moreover, electroculture is a fertile ground for innovation. As we delve deeper into this field, we can develop new tools and techniques that can amplify the benefits of electroculture. Each discovery, each innovation, brings us one step closer to harnessing the full potential of electroculture.

In conclusion, the future of electroculture is a tapestry woven with threads of challenges and opportunities. While the challenges are formidable, the opportunities are immense. As we continue our journey of discovery, we can look forward to a future where electroculture is not just a fringe concept, but a cornerstone of sustainable agriculture. A future where we harness the power of electricity to nurture life, to enhance plant growth, and to feed our world.

Conclusion

As we draw this enlightening journey to a close, it's essential to reflect on the transformative power of electroculture. This book has taken you through the fascinating intersection of electricity and biology, demonstrating how the invisible forces of electricity can be harnessed to enhance plant growth and productivity.

Throughout these pages, we've explored the historical development of electroculture, the science behind the process, and the practical applications in your garden or farm. We've delved into the various electroculture systems, the tools and materials needed, and the safety precautions to observe.

We've also highlighted the importance of continuous learning and experimentation in this field. Electroculture is not a static practice but an evolving one, constantly shaped by new discoveries and innovations. It's a field that invites curiosity, rewards patience, and celebrates the spirit of exploration.

However, it's important to remember that electroculture is not just about improving crop yields or enhancing plant health. It's about fostering a deeper connection with the natural world and gaining a greater appreciation for the intricate interplay between electricity and life. It's about seeing our gardens not just as spaces for growing plants, but as vibrant ecosystems teeming with unseen energy.

As you continue your journey in electroculture, remember that success lies not in the destination but in the journey itself. It's in the lessons learned, the challenges overcome, and the joy of witnessing the fruits of your labor.

In conclusion, electroculture offers a promising and exciting path forward for agriculture, one that harnesses the power of natural energies to create healthier, more vibrant gardens. As we look to the future, the potential of electroculture is vast and largely untapped. It's up to us to continue exploring, experimenting, and pushing the boundaries of what's possible.

Thank you for embarking on this journey with me. Here's to a future where our gardens are not just thriving, but truly electrifying!

JOHN FORD

Made in United States
Troutdale, OR
06/10/2024